Hybrid Graph Theory and Network Analysis

Cambridge Tracts in Theoretical Computer Science

Titles in the series

1. G. Chaitin *Algorithmic Information Theory*
2. L. C. Paulson *Logic and Computation*
3. M. Spivey *Understanding Z*
5. A. Ramsey *Formal Methods in Artificial Intelligence*
6. S. Vickers *Topology via Logic*
7. J.-Y. Girard, Y. Lafont & P. Taylor *Proofs and Types*
8. J. Clifford *Formal Semantics & Progmatics for Natural Language Processing*
9. M. Winslett *Updating Logical Databases*
10. K. McEvoy & J. V. Tucker (eds) *Theoretical Foundations of VLSI Design*
11. T. H. Tse *A Unifying Framework for Structured Analysis and Design Models*
12. G. Brewka *Nonmonotonic Reasoning*
14. S. G. Hoggar *Mathematics for Computer Graphics*
15. S. Dasgupta *Design Theory and Computer Science*
17. J. C. M. Baeten (ed) *Applications of Process Algebra*
18. J. C. M. Baeten & W. P. Weijland *Process Algebra*
19. M. Manzano *Extensions of First Order Logic*
21. D. A. Wolfram *The Clausal Theory of Types*
22. V. Stoltenberg-Hansen, I. Lindström & E. Griffor *Mathematical Theory of Domains*
23. E.-R. Olderog *Nets, Terms and Formulas*
26. P. D. Mosses *Action Semantics*
27. W. H. Hesselink *Programs, Recursion and Unbounded Choice*
28. P. Padawitz *Deductive and Declarative Programming*
29. P. Gärdenfors (ed) *Belief Revision*
30. M. Anthony & N. Biggs *Computational Learning Theory*
31. T. F. Melham *Higher Order Logic and Hardware Verification*
32. R. L. Carpenter *The Logic of Typed Feature Structures*
33. E. G. Manes *Predicate Transformer Semantics*
34. F. Nielson & H. R. Nielson *Two Level Functional Languages*
35. L. Feijs & H. Jonkers *Formal Specification and Design*
36. S. Mauw & G. J. Veltink (eds) *Algebraic Specification of Communication Protocols*
37. V. Stavridou *Formal Methods in Circuit Design*
38. N. Shankar *Metamathematics, Machines and Gödel's Proof*
39. J. B. Paris *The Uncertain Reasoner's Companion*
40. J. Dessel & J. Esparza *Free Choice Petri Nets*
41. J.-J. Ch. Meyer & W. van der Hoek *Epistemic Logic for AI and Computer Science*
42. J. R. Hindley *Basic Simple Type Theory*
43. A. Troelstra & H. Schwichtenberg *Basic Proof Theory*
44. J. Barwise & J. Seligman *Information Flow*
45. A. Asperti & S. Guerrini *The Optimal Implementation of Functional Programming Languages*
46. R. M. Amadio & P.-L. Curien *Domains and Lambda-Calculi*

Hybrid Graph Theory and Network Analysis

Ladislav Novak
University of Novi Sad

Alan Gibbons
University of Liverpool

CAMBRIDGE
UNIVERSITY PRESS

PUBLISHED BY THE PRESS SYNDICATE OF THE UNIVERSITY OF CAMBRIDGE
The Pitt Building, Trumpington Street, Cambridge, United Kingdom

CAMBRIDGE UNIVERSITY PRESS
The Edinburgh Building, Cambridge CB2 2RU, UK   www.cup.cam.ac.uk
40 West 20th Street, New York, NY 10011-4211, USA   www.cup.org
10 Stamford Road, Oakleigh, Melbourne 3166, Australia
Ruiz de Alarcón 13, 28014 Madrid, Spain

First published 1999

Printed in the United Kingdom at the University Press, Cambridge

Typeset in Monotype Times 10/13pt, in LaTeX [EPC]

A catalogue record of this book is available from the British Library

Library of Congress Cataloguing in Publication data

ISBN 0 521 46117 0 hardback

For our children, Ana, Chantal, Gabrielle, Rosalind and Uglješa

Contents

Preface   ix

1   Two Dual Structures of a Graph   1
   1.1   Basic concepts of graphs   1
   1.2   Cuts and circs   5
   1.3   Cut and circ spaces   9
   1.4   Relationships between cut and circ spaces   12
   1.5   Edge-separators and connectivity   16
   1.6   Equivalence relations among graphs   19
   1.7   Directed graphs   22
   1.8   Networks and multiports   25
   1.9   Kirchhoff's laws   33
   1.10   Bibliographic notes   38

2   Independence Structures   39
   2.1   The graphoidal point of view   39
   2.2   Independent collections of circs and cuts   42
   2.3   Maximal circless and cutless sets   48
   2.4   Circ and cut vector spaces   55
   2.5   Binary graphoids and their representations   62
   2.6   Orientable binary graphoids and Kirchhoff's laws   71
   2.7   Mesh and nodal analysis   76
   2.8   Bibliographic notes   79

3   Basoids   80
   3.1   Preliminaries   80
   3.2   Basoids of graphs   81
   3.3   Transitions from one basoid to another   87
   3.4   Minor with respect to a basoid   91
   3.5   Principal sequence   94
   3.6   Principal minor and principal partition   99
   3.7   Hybrid rank and basic pairs of subsets   102
   3.8   Hybrid analysis of networks   105

3.9   Procedure for finding an optimal basic pair 109
3.10 Bibliographic notes 113

4 Pairs of Trees   115
4.1   Diameter of a tree 115
4.2   Perfect pairs of trees 119
4.3   Basoids and perfect pairs of trees 126
4.4   Superperfect pairs of trees 130
4.5   Unique solvability of affine networks 134
4.6   Bibliographic notes 139

5 Maximally Distant Pairs of Trees   141
5.1   Preliminaries 141
5.2   Minor with respect to a pair of trees 145
5.3   Principal sequence 149
5.4   The principal minor 155
5.5   Hybrid pre-rank and the principal minor 159
5.6   Principal partition and Shannon's game 164
5.7   Bibliographic notes 167

Bibliography   168

Index   174

Preface

This research monograph is concerned with two dual structures in graphs. These structures, one based on the concept of a circuit and the other on the concept of a cutset are strongly interdependent and constitute a hybrid structure called a graphoid. This approach to graph theory dealing with graphoidal structures we call *hybrid graph theory*. A large proportion of our material is either new or is interpreted from a fresh viewpoint. Hybrid graph theory has particular relevance to the analysis of (lumped) systems of which we might take electrical networks as the archetype. Electrical network analysis was one of the earliest areas of application of graph theory and it was essentially out of developments in that area that hybrid graph theory evolved. The theory emphasises the duality of the circuit and cutset spaces and is essentially a vertex independent view of graphs. In this view, a circuit or a cutset is a subset of the edges of a graph without reference to the endpoints of the edges. This naturally leads to working in the domain of graphoids which are a generalisation of graphs. In fact, two graphs have the same graphoid if they are 2-isomorphic and this is equivalent to saying that both graphs (within a one-to-one correspondence of edges) have the same set of circuits and cutsets.

Historically, the study of hybrid aspects of graphs owes much to the foundational work of Japanese researchers dating from the late 1960's. Here we omit the names of individual researchers, but they may be readily identified through our bibliographic notes.

First two chapters could be seen as a bridge between traditional graph theory and the graphoidal perspective. They may be also taken as a kind of separate subtext. The proofs to be found in these pages are usually different from those of the traditional literature. The central bulk (Chapers 3, 4, and 5) of the text is then pure hybrid graph theory. It may be noted that the presentation of this material is such that it is easily viewed from a matroidal perspective. The last sections of Chapters 1 to 4 are devoted to fundamental aspects of network analysis in a form which is rather formal. Such an approach could be suitable for readers who appreciate an abstract and formal point of view in traditional engineering texts such as engineering theorists, computer scientists and mathematicians.

The body of theorems and propositions developed in the text provides a solid basis for the correctness of algorithms presented. Although the question of their complexity has not been addressed in detail, wherever the quantity of output is polynomially bound, the algorithms clearly run in low order polynomial time, mostly involving elementary graph operations such as the removal or contraction of edges of the graph.

We have adopted the practice of only referring to sources of material within the Bibliographic notes which are appended to each chapter. Our mathematical notation is generally standard although we have adopted the local convention of using bold face type to denote *sets of sets* and plain font for other sets.

We thank the University of Warwick, Department of Computer Science and the University of Novi Sad, Department of Electrical Engineering and Computing for providing facilities and the British Council and British Royal Society who partially funded this work by supporting one of us (LN) during the course of it.

We also thank our students and technical staff, particularly Martyn Amos, Craig Eales, Ken Chan and Andrew Mellor who provided valuable technical support in the preparation of the text. Special thanks are due to Boris Antić and Ida Pu who generously gave of their time to create final versions of all our figures.

Finally, we express our gratitude to CUP and their staff, in particular David Tranah for their patience and support.

Ladislav Novak
Alan Gibbons

1

Two Dual Structures of a Graph

The concept of graph inherently includes two dual structures, one based on circuits and the other based on cutsets. These two structures are strongly interdependent. They constitute the so-called graphoidal structure which is a deep generalization of the concept of graph and comes from matroid theory. The main difference between graph and graphoid is that the latter requires no concept of vertex while the first presumes vertex as a primary notion. After some bridging material, our approach will be entirely vertex-independent. We shall concentrate on the hybrid aspects of graphs that naturally involve both dual structures. Because of this approach, the material of this text is located somewhere between graph theory and the theory of matroids. This enables us to combine the advantages of both an intuitive view of graphs and formal mathematical tools from matroid theory.

Starting with the classical definition of a graph in terms of vertices and edges we define circs and cuts and then circuits and cutsets. In this context, a circuit (respectively, cutset) is a minimal circ (cut) in the sense that no proper subset of it is also a circ (cut). We consider some collective algebraic properties and mutual relationships of circs and cuts. Vertex and edge-separators are introduced and through these we define various kinds of connectivity. The immediate thrust is towards a vertex-independent description of graphs, so that later all theorems and propositions will be vertex-independent.

The last two sections of the chapter are devoted to the notions of multiports and Kirchhoff's laws which are basic concepts in network analysis.

1.1 Basic concepts of graphs

A *graph* G is a triple $(V, E; f)$ where $V \neq \emptyset$ and E are disjoint finite sets, and $f : E \to 2^V$, is an *incidence map* from E to the power set of V such that $1 \leq |f(e)| \leq 2$, for all $e \in E$. The elements of V are called *vertices*, and the elements of E are called *edges*. Generally, the map f is not a surjection, that is, $f(E) \subseteq V$. If $f(E) \subset V$, then the set difference $V \setminus f(E)$ is nonempty and its members are called *isolated vertices*. If the upper bound $|f(e)| \leq 2$ is omitted in the above definition, the triple $(V, E; f)$ is called a *hypergraph*. The map f describes how edges in E connects vertices in V. If an edge e connects a pair of distinct vertices, that is, if $|f(e)| = 2$, then e is a *regular edge*; if $|f(e)| = 1$ it is a *loop edge*; otherwise, that is if $|f(e)| > 2$ then e is a *hyper edge*. The map $f^* : V \to 2^E$ defined as $f^*(v) = \{e \in E | v \in f(e)\}$ is called the *transpose* of f. Since $v \in f(e)$ iff $e \in f^*(v)$ we conclude that $f^{**}(e) = \{v \in V \mid e \in f^*(v)\} = f(e)$ for all $e \in E$. If $e \in f^*(v)$ or $v \in f(e)$ then we say that edge e is *incident* with vertex v, and vice versa. Two edges are *adjacent* if they are incident with the same vertex. The map f provides that each edge is incident with not less than one and not more than two vertices. Two edges e' and e'' of a graph are said to be *parallel* edges of the graph if $f(e') = f(e'')$. Obviously the relation 'to be parallel' is symmetric and transitive. The number of regular edges incident with a vertex of a graph plus twice the number of loop-edges attached to the same vertex, is called the *degree* of that vertex. Clearly, a vertex of a graph is an isolated vertex iff its degree is zero.

Let $G = (V, E; f)$ be the graph where $V = \{v_1, \ldots, v_8\}$, $E = \{e_1, \ldots, e_{10}\}$ and f be the map described by: $e_1 \to \{v_1\}$, $e_2 \to \{v_1, v_2\}$, $e_3 \to \{v_1, v_2\}$, $e_4 \to \{v_1, v_3\}$, $e_5 \to \{v_2, v_3\}$, $e_6 \to \{v_2, v_4\}$, $e_7 \to \{v_3, v_4\}$, $e_8 \to \{v_3, v_5\}$, $e_9 \to \{v_7, v_8\}$, $e_{10} \to \{v_7, v_8\}$. Figure 1.1 is an illustration of G and we see that vertex v_6 is an isolated vertex, edges e_2 and e_3 are parallel edges as are edges e_9 and e_{10}, edge e_1 is a loop-edge while all other edges are regular. Edge e_1 is incident with vertex v_1, edge e_4 is incident with v_1 and v_3 while, for example, e_2 and e_3 are both incident with vertices v_1 and v_2. Notice that $f^*(v_1) = \{e_1, e_2, e_3, e_4\}$, $f^*(v_5) = \{e_8\}$, $f^*(v_6) = \emptyset$.

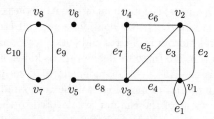

Figure 1.1. An example of a graph.

Given a graph $G = (V, E; f)$ let $e \in E$ be an arbitrary edge and let $v_1, v_2 \in V$, be a pair of vertices such that $f(e) = \{v_1, v_2\}$. Then we can define the following three operations on G related to e:

- *Removal* of e from G produces the new graph $\overline{G} = (\overline{V}, \overline{E}; \overline{f})$ where $\overline{V} = V$, $\overline{E} = E \setminus \{e\}$ and $\overline{f}$ is the restriction of f from E to $\overline{E} = E \setminus \{e\}$.

- *Short-circuiting* of e from G, providing that vertices v_1 and v_2 are distinct, produces the new graph $\underline{G} = (\underline{V}, \underline{E}; \underline{f})$ where $\underline{V} = (V \backslash \{v_1, v_2\}) \cup \{v_0\}$, $\underline{E} = E$ and $\underline{f}$ is a map from $\underline{E}$ to $\underline{V}$ such that $\underline{f}^*(v) = f^*(v)$ for all $v \in V \backslash \{v_1, v_2\}$ and $\underline{f}^*(v_0) = f^*(v_1) \cup f^*(v_2)$.

- *Contraction* of $\bar{e}$ can be described as consecutive applications of the operations short-circuiting of $\bar{e}$ and removal of $\bar{e}$.

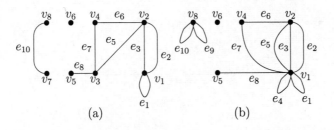

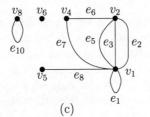

Figure 1.2. Illustrating (a) edge removal (b) edge shortcircuiting and (c) edge contraction operations.

Figure 1.2(a) shows the graph obtained from Figure 1.1 by removing the edges e_4 and e_9, Figure 1.2(b) shows the graph obtained from Figure 1.1 by short-circuiting the edges e_4 and e_9, whereas Figure 1.2(c) shows the graph obtained from Figure 1.1 by contracting e_4, and e_9.

Given a graph G with edge set E, let $S \subseteq E$. The *reduction* of G from E to S is the graph $G|S$ obtained from G by removing all the edges in $E \backslash S$. The *contraction* of G from E to S, is the graph $G \cdot S$ obtained from G by contracting the edges that belong to $E \backslash S$.

A graph $G_0 = (V_0, E_0; f_0)$ is said to be a *subgraph* of a graph $G = (V, E; f)$ if $V_0 \subseteq V$, $E_0 \subseteq E$, $f(E_0) \subseteq V_0$ and f_0 is the restriction of f from E to E_0. Let $G = (V, E; f)$ and let V' be a subset of V. The graph $G' = (V', E'; f')$ is said to be a *vertex-induced subgraph* of G if E' contains all edges in E that join pairs of vertices in V' only, that is, if $f^*(V') = E'$. It is clear that V' uniquely specifies the subgraph G'. Let $G = (V, E; f)$ and let E'' be a subset of E. Then the graph $G'' = (V'', E''; f'')$ is said to be an *edge-induced subgraph* of G if V'' contains all vertices of V that are joined by edges in E'' only, that is, if $f(E'') = V''$. It is clear that E'' uniquely specifies the subgraph G''.

As an example, the graph of Figure 1.3(a) (respectively, Figure 1.3(b) is a vertex-induced (an edge-induced) subgraph of the graph of Figure 1.1 associated with the vertex subset $V' = \{v_1, v_2, v_3, v_5, v_8\}$ (edge subset $E'' = \{e_1, e_2, e_5, e_8\}$).

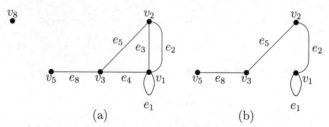

(a)   (b)

Figure 1.3. (a) With respect to Figure 1.1 (a) is a vertex-induced subgraph and (b) is an edge-induced subgraph.

Let G be a graph with vertex set V and edge set E. The edge-induced subgraph of G associated with E can be obtained from G by removing all isolated vertices. In a similar way, given an edge subset E', the edge-induced subgraph G' of a graph G associated with E' and the graph G'' obtained by removing all edges in $E\backslash E'$ differ only with respect to isolated vertices. In fact, the subgraph G' is obtained from the subgraph G'' by removing all isolated vertices.

A subset V' of the vertex set V of a graph G is a *vertex-separator* of G if each edge of G either joins a pair of vertices in V' or joins a pair of vertices in $V\backslash V'$. Clearly, the empty set and the set V itself are trivial vertex-separators. It is easy to see that the union or intersection of vertex-separators is a vertex-separator and the complement of a vertex-separator is also a vertex-separator. A vertex-separator is an *elementary vertex-separator* if it contains no vertex-separator as a proper subset.

Proposition 1.1 *Let G be a graph with the vertex set V and let V' and $V'' \subset V'$ be two vertex-separators of G. Then $V'\backslash V''$ is also a vertex-separator of G.*

Proof Let V' and $V'' \subset V'$ be two vertex separators of G. Since $V'' \subset V'$ and V'' is a vertex-separator it follows that no edge joins a vertex from V'' with a vertex from V'. But V' is also a vertex-separator and hence each edge of G either joins a pair of vertices in $V' \setminus V''$ or joins a pair of vertices in V'' or joins a pair of vertices in $V\backslash V'$. Consequently each edge of G either joins a pair of vertices in $V' \setminus V''$ or joins a pair of vertices in $V\backslash(V' \setminus V'')$.   □

Corollary 1.2 [to Proposition 1.1] *Any vertex-separator of a graph G is either an elementary vertex-separator of G or a union of disjoint elementary vertex-separators of G.*

Thus, the elementary vertex-separators of G are disjoint nonempty subsets of V whose union is V.

Let V' be a proper nonempty subset of the vertex set V and let G' and G'' be the vertex-induced subgraphs associated with V' and $V'' = V \backslash V'$ respectively. Denote by E' and E'' the edge sets associated with G' and G''. Then V' is a vertex-separator iff $E' \cup E'' = E$.

A graph G is said to be *1-connected* (or simply *connected*) if it has no vertex-separator other than the empty set and the set V. A graph which is not 1-connected is called *unconnected*. Let the set of all elementary vertex-separators of a graph G be $\{V_1, \ldots, V_q\}$ and let G_k be the vertex-induced subgraph associated with V_k. Obviously for each $k = 1, \ldots, q$, G_k is 1-connected. We say that $G_k, k = 1, \ldots, q$ are *1-connected components* of G.

1.2 Cuts and circs

Let V' be a nonempty proper subset of the vertex set V of a graph G. The set of all edges of the graph that join vertices of V' with vertices in $V \backslash V'$ is called a *cut* generated by the vertex subset V'. Notice that V' and $V \backslash V'$ both define the same cut. Notice also that any collection of isolated vertices defines a trivial (empty) cut.

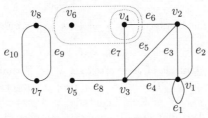

Figure 1.4. Two distinct vertex subsets indicating the same cut of a graph.

A nonempty proper subset V' of nonisolated vertices of a graph may also define an empty cut, even if its complement contains nonisolated vertices. See for example the subset $\{v_7, v_8\}$ of vertices of the graph in Figure 1.4. Clearly, *the cut defined by a nonempty proper vertex subset V' is empty iff V' is a vertex-separator.*

For a given subset V' of the vertex set V of a graph G there exists a unique cut S. The converse statement is generally not true. For a given cut S of a graph, there may exist more than one vertex subset generating S. This is illustrated for example in Figure 1.4 where two distinct vertex subsets $\{v_4\}$ and $\{v_4, v_6\}$ indicate the same cut $\{e_6, e_7\}$.

A nonempty cut of a graph is said to be a *cutset* if it is a minimal cut, in the sense that it contains no other nonempty cut as a proper subset. Accordingly, no proper subset of a cutset is a cutset. Also, no cutset is empty. A cut consisting of only one edge is called a *self-cutset*.

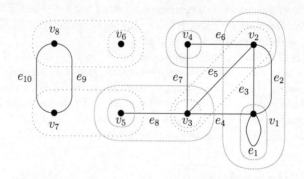

Figure 1.5. Examples of cuts and cutsets of a graph.

For the graph of Figure 1.1, several examples of cuts are shown in Figure 1.5. Closed dotted curves define different nonempty proper subsets of the vertex set V, each uniquely defining a cut consisting of the edges crossed by the closed dotted curve. Sparsely-dotted closed curves indicate nonminimal cuts while the rest indicate cutsets.

Proposition 1.3 *Let S' and $S'' \subset S'$ be cuts of a graph G. Then $S'\backslash S''$ is also a cut of G.*

Proof Let V' be a subset of the vertex set V of G that generates the cut S' and suppose S' contains the cut S'' as a proper subset. Then obviously at least one of the subsets V' or $V \setminus V'$ contains *as a proper subset* a subset V'' that generates S''. Without loss of generality suppose that $V'' \subset V'$. Then no edge of G connects a vertex from V'' with a vertex from $V' \setminus V''$ and therefore every edge in $S'\backslash S''$ connects a vertex from $V' \setminus V''$ with a vertex in $V \setminus (V' \setminus V'')$. Therefore, $S'\backslash S''$ is a cut of G. □

Corollary 1.4 [to Proposition 1.3] *Any cut of a graph is either a cutset or a union of disjoint cutsets of the graph.*

To illustrate Corollary 1.4 consider Figure 1.5. The cut $\{e_2, e_3, e_4, e_6, e_7, e_8\}$, defined by the vertex subset $\{v_2, v_3\}$, is the union of three disjoint cutsets $\{e_2, e_3, e_4\}, \{e_8\}$ and $\{e_6, e_7\}$.

The cut generated by a vertex subset which is of cardinality one is called a *star*. Obviously, a star consists of all regular edges incident to a particular vertex. Loops do not contribute to stars. Each star of a graph is a cut and therefore it is either a cutset of the graph or the union of disjoint cutsets of the graph. When a star contains a cutset as a proper subset, we say it is a *separable star*. A graph is said to be *2-connected* if it is 1-connected and has no separable stars.

An edge subset of a graph G is defined to be a *circ* if every vertex of the associated edge-induced subgraph is of even degree. Clearly an empty set of edges is also a circ. A nonempty circ is called a *circuit* if it is a minimal circ in the sense that it contains no other nonempty circ as a proper subset. Accordingly, no proper subset of a circuit is a circuit. Also, no circuit is empty. A circ which consists of only one edge we shall call a *self-circuit*. Obviously, an edge forms a self-circuit iff it is a loop. For the graph of Figure 1.6(a), the following subsets of the edge set E of the graph are circuits: $\{e_1\}$, $\{e_2, e_3\}$, $\{e_9, e_{10}\}$, $\{e_5, e_6, e_7\}$, $\{e_2, e_4, e_6, e_7\}$, $\{e_2, e_4, e_5\}$, $\{e_3, e_4, e_5\}$ and $\{e_3, e_4, e_6, e_7\}$. Figure 1.6(a) illustrates a nonminimal circ (solid bold lines) and Figure 1.6(b) shows a circuit (dashed bold lines). Nonminimal circs are also $\{e_2, e_3, e_5, e_6, e_7\}$, $\{e_1, e_2, e_3\}$, $\{e_1, e_5, e_6, e_7\}$ and $\{e_1, e_2, e_3, e_5, e_6, e_7\}$.

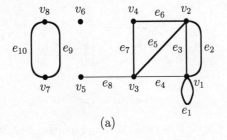

(a)

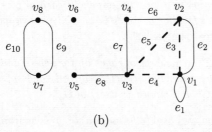

(b)

Figure 1.6. (a) A nonminimal circ. (b) A circuit.

If an edge subset of a graph is a circuit then, in the associated edge-induced subgraph, each vertex has degree equal to 2. The converse is generally not true. For example for the graph of Figure 1.1 the edge-induced subgraph associated with the circ $\{e_1, e_5, e_6, e_7\}$ contains vertices that are all of degree 2 only, but nevertheless it is not a minimal circ.

Proposition 1.5 *Let C_o and $C \subset C_o$ be circs of a graph G. Then $C \backslash C_o$ is also a circ of the graph.*

Proof Suppose the circ C has a circ C_o as a proper subset. By definition, vertices in each of the associated edge-induced subgraphs have even degrees. Because the difference of two even numbers is also even, each vertex in the edge-induced subgraph associated with $C \backslash C_o$ also has even degree.   □

Corollary 1.6 [to Proposition 1.5] *Any circ of a graph is either a circuit or a union of disjoint circuits of the graph.*

As an illustration observe that in Figure 1.6(a) circ $\{e_1, e_2, e_3, e_5, e_6, e_7, e_9, e_{10}\}$ is the union of the following four disjoint circuits: $\{e_1\}, \{e_2, e_3\}, \{e_9, e_{10}\}$ and $\{e_5, e_6, e_7\}$.

The following two types of graphs are of particular importance: a graph G with edge set E is *bipartite* if E is a cut of G; it is *Eulerian (cobipartite)* if E is a circ of G. Figure 1.7 shows a graph that is both bipartite and Eulerian.

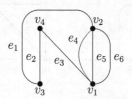

Figure 1.7. A graph that is both bipartite and Eulerian.

Remark 1.7 *A graph is a bipartite graph iff each circ is of even cardinality* (see Lemma 1.24). *Dually, a graph is an Eulerian graph iff each cut is of even cardinality* (see Lemma 1.25).

Lemma 1.8 *Let E' be a subset of edges of a graph G and let G' be the graph obtained from G by removing (contracting) all edges of $E \backslash E'$. Then, each circuit (cutset) of G' is a circuit (cutset) of G.*

Proof Observe that removal (contraction) of an edge from a graph G destroy all circuits (cutsets) of G that contain this edge and leave all other circuits (cutsets) unchanged. Therefore each circuit (cutset) of the graph G', obtained from G by removing (contracting) all edges of $E \backslash E'$, is a circuit (cutset) of G.   □

Lemma 1.9 *Let E' be a subset of the edge set E of a graph G and let G' be a graph obtained from G by contracting (removing) all edges of $E\backslash E'$. Then, a subset C' (S') of E' is a circuit (cutset) of G' iff there exists a circuit C (cutset S) of G such that $C' = E' \cap C$ $(S' = E' \cap S)$.*

Proof Observe that contraction (removal) of an edge $e \in E$ from a graph G transforms a circuit C (cutset S) of G to the circuit $C' = C \setminus \{e\}$ (cutset $S' = S\backslash\{e\}$) of G' if $e \in C$ ($e \in S$) and keeps a circuit C (cutset S) unchanged, otherwise. Therefore after contracting (removing) all edges of $E\backslash E'$ from G every circuit C (cutset S) of G transforms into a circuit $C' = E' \cap C$ (cutset $S' = E' \cap S$) of G'. On the other hand, by construction, there are no circuits (cutsets) of G' other then those obtained by contracting (removing) the edges of $E\backslash E'$ from G. □

Notice that Lemmas 1.8 and 1.9 remain valid if we replace cutset with cut and circuit with circ.

1.3 Cut and circ spaces

In what follows, we shall frequently use the binary set operation $\oplus$ called the *ring sum* $A \oplus B \stackrel{\text{def}}{=} (A\backslash B) \cup (B\backslash A)$. An equivalent definition is $A \oplus B = (A \cup B)\backslash(B \cap A)$. It is easy to see that $\oplus$ is both an associative and a commutative operation. The following obviously hold: $B \oplus \emptyset = B, \emptyset \oplus B = B$ and $B \oplus B = \emptyset$. Thus, the empty set is an identity element for the operation $\oplus$ and every set is its own inverse.

Let G be a graph and let the set $\mathbf{C}(G)$ be the set of all circs and let the set $\mathbf{S}(G)$ be the set of all cuts of the graph G. We shall show that $\mathbf{C}(G)$ and $\mathbf{S}(G)$ are both closed with respect to $\oplus$.

Lemma 1.10 *Given a nonempty proper subset V' of the vertex set of a graph, of cardinality p, let S be the cut associated with V'. Then $S = S'_1 \oplus \cdots \oplus S'_p$, where $S'_1, \ldots, S'_p$ are the stars of vertices in V'.*

Proof Clearly, each star of a graph is a cut. Let S' be the ring sum of the stars of all vertices in V'. Because each edge of the cut S appears in exactly one star associated with a vertex in V', we conclude that S' contains S. On the other hand, the edges that join vertices in V' are not contained in S'. Thus, $S' = S$. □

The following proposition is an obvious consequence of Lemma 1.10:

Proposition 1.11 *The ring sum of any two cuts of a graph is also a cut.*

As an illustration, consider the graph of Figure 1.1 and the cuts $S_1 = \{e_7, e_6\}$ and $S_2 = \{e_7, e_5, e_3, e_2\}$. By inspection, $S_1 \oplus S_2 = \{e_6, e_5, e_3, e_2\}$ which is also a cut of the graph. Because of Proposition 1.11 the set of all cuts of a graph we call a *cut space*.

Corollary 1.12 [to Proposition 1.11 and Corollary 1.4] *The ring sum of two cutsets is either a cutset or the union of a number of disjoint cutsets.*

For example, for the graph of Figure 1.1 the ring sum of cutsets $\{e_2, e_3, e_5, e_7\}$ and $\{e_4, e_5, e_6\}$, that is, $\{e_2, e_3, e_4, e_6, e_7\}$ is not a cutset but is the union of the disjoint cutsets $\{e_2, e_3, e_4\}$ and $\{e_6, e_7\}$.

Corollary 1.13 [to Proposition 1.11] *Let $\mathbf{S_o}(G)$ be the set of all cutsets of a graph. Then the following two conditions hold:*
    (S1) *No proper subset of a member of $\mathbf{S_o}(G)$ is itself a member of $\mathbf{S_o}(G)$.*
    (S2) *If S_1 and S_2 are distinct members of $\mathbf{S_o}(G)$ and $x \in S_1 \cap S_2$ then for each $y \in S_1 \backslash S_2$ there exists a member S of $\mathbf{S_o}(G)$ such that $y \in S \subseteq (S_1 \cup S_2) \backslash \{x\}$.*

Proof (S1) Follows immediately from the minimality of cutsets. (S2) Let S_1 and S_2 be two distinct members of $\mathbf{S_o}(G)$ and let $x \in S_1 \cap S_2$. If $y \in S_1 \backslash S_2$ then $y \in S_1 \oplus S_2$. According to Proposition 1.11, $S_1 \oplus S_2 \in \mathbf{S_o}(G)$ and hence there exists a member S of $\mathbf{S_o}(G)$ such that $y \in S \subseteq S_1 \oplus S_2$. Because $S_1 \oplus S_2$ contains all elements in $S_1 \cup S_2$ and does not contain elements in $S_1 \cap S_2$, we conclude that $y \in S \subseteq (S_1 \cup S_2) \backslash \{x\}$. $\square$

Lemma 1.14 *The ring sum of two cuts of even cardinality also has even cardinality.*

Proof Let S_1 and S_2 be cuts of even cardinality, respectively $2k_1$ and $2k_2$. Then
$$\begin{aligned} |S_1 \oplus S_2| &= |S_1| + |S_2| - 2|S_1 \cap S_2| = 2k_1 + 2k_2 - 2|S_1 \cap S_2| \\ &= 2(k_1 + k_2 - |S_1 \cap S_2|). \end{aligned}$$
$\square$

Lemma 1.15 *For $j = 0, 1, 2$ let $G_j = (V, E_j, f_j)$, be the restriction of a graph $G = (V, E, f)$ from E to E_j and let $S_j(v)$ be the star associated with a vertex v in G_j. If $E_0 = E_1 \oplus E_2$ then $S_0(v) = S_1(v) \oplus S_2(v)$ for every vertex v of G.*

Proof It is sufficient to observe that all restricted graphs $G_0 = (V, E_0, f_0)$, $G_2 = (V, E_1, f_1)$ and $G_3 = (V, E_2, f_2)$ have the same vertex set V and that for all $v \in V$ an edge belongs to E_j iff it belongs to $S_j(v)$. $\square$

Proposition 1.16 *The ring sum of any two circs of a graph is also a circ.*

Proof Let C_1 and C_2 be two circs of a graph G and let $C_0 = C_1 \oplus C_2$. By $G_j = (V, C_j, f_j)$, $j = 0, 1, 2$ we denote the restriction of $G = (V, E, f)$ from E to C_j. To prove that C_0 is a circ it is enough to show that each vertex in G_0 has even degree. Let v be a vertex of G_0 and for $j = 0, 1, 2$, let $S_j(v)$ and $d_j(v)$ denote the cut and the degree associated with v in G_j respectively. From Lemma 1.15, $S_0(v) = S_1(v) \oplus S_2(v)$ and hence:
$$|S_0(v)| = |S_1(v) \oplus S_2(v)| = |S_1(v)| + |S_2(v)| - 2|S_1(v) \cap S_2(v)|.$$
Now, $d_i(v) = |S_i(v)| + 2k_i$, where k_i is the number of loops of v and so:
$$(d_0(v) - 2k_0) = (d_1(v) - 2k_1) + (d_2(v) - 2k_2) - 2|S_1(v) \cap S_2(v)|.$$
Consequently, $d_0(v) = d_1(v) + d_2(v) + 2k_0 - 2k_2 - 2k_1 - 2|S_1(v) \cap S_2(v)|$
$= d_1(v) + d_2(v) + 2(k_0 - k_2 - k_1 - |S_1(v) \cap S_2(v)|)$.

Because C_1 and C_2 are circs, $d_1(v)$ and $d_2(v)$ are even and therefore so is $d_0(v)$. It follows that C_0 is a circ. $\square$

As an illustration consider the graph of Figure 1.1 with the two circs $C_1 = \{e_2, e_3, e_5, e_6, e_7\}$ and $C_2 = \{e_2, e_4, e_5\}$. Now $C_1 \oplus C_2 = \{e_3, e_4, e_6, e_7\}$ which also is a circ of the graph. Because of Proposition 1.16 we call the set of all circs of a graph a *circ space*.

Corollary 1.17 [to Proposition 1.16 and Corollary 1.6] *The ring sum of two circuits of a graph is either a circuit or the union of a number of disjoint circuits.*

For example, for the graph of Figure 1.1, the ring sum of the circuits $C_1 = \{e_2, e_4, e_5\}$ and $C_2 = \{e_3, e_4, e_6, e_7\}$ is $C_3 = \{e_2, e_3, e_5, e_6, e_7\}$ which is not a circuit but is the union of the disjoint circuits $\{e_2, e_3\}$ and $\{e_5, e_6, e_7\}$. On the other hand, the ring sum of two nonminimal circs or a nonminimal circ and a circuit might be a circuit. For example, $C_1 \oplus C_3 = C_2$ is a circuit even though C_3 is not.

Corollary 1.18 [to Proposition 1.16] *Let $\mathbf{C}_o(G)$ be the set of all circuits of a graph G. Then the following two conditions hold:*

(C1) *No proper subset of a member of $\mathbf{C}_o(G)$ is itself a member of $\mathbf{C}_o(G)$.*

(C2) *If C_1 and C_2 are distinct members of $\mathbf{C}_o(G)$ and $x \in C_1 \cap C_2$ then for each $y \in C_1 \backslash C_2$ there exists a member C of $\mathbf{C}_o(G)$ such that $y \in C \subseteq (C_1 \cup C_2) \backslash \{x\}$.*

Proof (C1) Follows immediately from the minimality of cutsets. (C2) Let C_1 and C_2 be two distinct members of $\mathbf{C}_o(G)$ and let $x \in C_1 \cap C_2$. If $y \in C_1 \backslash C_2$ then $y \in C_1 \oplus C_2$. According to Proposition 1.16 $C_1 \oplus C_2 \in \mathbf{C}_o(G)$ and hence there exists a member C of $\mathbf{C}_o(G)$ such that $y \in C \subseteq C_1 \oplus C_2$. Because $C_1 \oplus C_2$ contains all elements in $C_1 \cup C_2$ and does not contain the elements in $C_1 \cap C_2$, we conclude that $y \in C \subseteq (C_1 \cup C_2) \backslash \{x\}$. $\square$

Remark 1.19 It is easy to see that the following seemingly stronger version of (C2) is also true:

(C2)* *For any two distinct circuits C_1 and C_2 of a graph and any two edges x and y (distinct or not) of $C_1 \cap C_2$, there is a circuit $C \subseteq (C_1 \cup C_2)\backslash\{x,y\}$.*

1.4   Relationships between cut and circ spaces

The space $\mathbf{C}(G)$ of all circs and the space $\mathbf{S}(G)$ of all cuts of a graph are interdependent. This section describes this dependency.

Proposition 1.20 [Orthogonality Theorem] *Let S be a cut and let C be a circ of a graph G. Then the cardinality of $C \cap S$ is even.*

Proof   Let $G_c = G|C$ be the restriction of G from E to C and let V_c be the associated vertex set of G_c. From Lemma 1.9, the intersection $S \cap C$, which we shall denote by S_c, is a cut of G_c. Let V_s be a vertex-subset of V_c that generates S_c. According to Lemma 1.10, the cut S_c can be expressed as $S_1^c \oplus \cdots \oplus S_p^c$, where $S_1^c, \ldots, S_p^c$ are the stars of all the vertices in V_s. By definition of circ, each vertex of G_c is of even degree and hence each star S_k^c, $k = 1, \ldots, p$, has even cardinality. Each star is a cut and it follows from Proposition 1.11 and Lemma 1.14 that $S_c = S \cap C$ is also of even cardinality. □

Remark 1.21   Because each cutset is a cut and each circuit is a circ, the statement of Proposition 1.20 remains true if we replace circ with circuit and cut with cutset.

Two collections $\mathbf{A}$ and $\mathbf{B}$ of subsets of a finite set E are said to be *orthogonal* if for each $A \in \mathbf{A}$ and each $B \in \mathbf{B}$, $|A \cap B|(modulo\ 2)=0$. The following corollary of Proposition 1.20 justifies the name Orthogonality Theorem:

Corollary 1.22 [of Proposition 1.20] *The set of all circs $\mathbf{C}(G)$ and the set of all cuts $\mathbf{S}(G)$ of a graph G are orthogonal.*

The following statement is equivalent to Proposition 1.20: *If a subset of edges of a graph is a cut (respectively, circ) then it has an even number of edges in common with every circ (cut) of the graph.* The converse of this statement is also true. The following proposition combines this reformulation of Proposition 1.20 and its converse in a single iff statement.

Proposition 1.23 *A subset of edges of a graph is a cut (circ) iff it has an even number of common edges with every circ (cut) of the graph.*

In order to prove Proposition 1.23, we first prove the following two lemmas.

Lemma 1.24 *A graph is bipartite iff every circ is of even cardinality.*

Proof Let C be a circ of a bipartite graph G. Because the whole edge set E of G is a cut, we have from the Orthogonality Theorem (Proposition 1.20) that $E \cap C$ is of even cardinality. But C is a proper subset of E, that is $E \cap C = C$, and so C is of even cardinality.

Conversely, let us assume that each circ is of even cardinality. Consider a binary relation on the vertex set V defined as follows: two vertices (v', v''), not necessarily distinct, are related iff there is a path of even length between v' and v''. A path between v' and v'' of length k is a sequence of (not necessarily distinct) vertices $v' = v_0, v_1, \ldots, v_k = v''$ in which every pair of consecutive vertices corresponds to an edge (either a regular edge or a self-circuit). Because each circ is of even cardinality, if there is a path of even length between two vertices, then all paths between this pair of vertices are also of even length. Hence, this relation is well defined. The relation is also easily seen to be reflexive, symmetric and transitive and hence is an equivalence relation. Clearly, the relation induces a bipartition of the vertex set V such that each edge of the graph joins vertices that belong to different equivalence classes. Hence the complete edge set forms a cut of the graph which consequently is bipartite. □

Lemma 1.25 *A graph is Eulerian iff each cut is of even cardinality.*

Proof Let G be an Eulerian graph and let S be a cut generated by the subset V' of the vertex set V. By definition of Eulerian, each star of G has even cardinality. From Lemma 1.10, S can be expressed as the ring sum of the stars associated with the vertices in V' or in $V \backslash V'$. Because all the stars are of even cardinality, it follows from Lemma 1.14, that $|S|$ is also even. Thus, any cut of an Eulerian graph is of even cardinality.

Conversely, if each cut of G is of even cardinality then, in particular, so is each star. It follows that the degree of each vertex is even and so G is Eulerian. □

Proof (of Proposition 1.23) Let a subset of edges of a graph be a cut (circ). Then from the Orthogonality Theorem (Proposition 1.20) it has an even number of common edges with every circ (cut) of the graph.

Conversely, suppose that a subset A (respectively, B) of edges of a graph G has an even number of edges in common with every circ (cut). Let $G \cdot A$ $(G \mid B)$ be the contraction (restriction) of $G = (V, E)$ from E to A (B) and let C' (S') be a circ (cut) of $G \cdot A$ ($G \mid B$). From Lemma 1.9 there exists a circ C (cut S) of G such that $C' = A \cap C$ ($S' = B \cap S$). Since the subset A (B) has an even number of edges in common with every circ C (cut S) of the graph G it follows that $C' = A \cap C$ ($S' = B \cap S$) is of even cardinality. Thus, using Lemma 1.24 (respectively, Lemma 1.25) we have proved that $G \cdot A$ ($G \mid B$) is bipartite (Eulerian) and therefore A (B) is a cut (circ) of $G \cdot A$ $(G \mid B)$. But, according to Lemma 1.8, every cut (circ) of $G \cdot A$ $(G \mid B)$ is also a cut (circ) of G, which completes the proof. □

From the Orthogonality Theorem we have:

Proposition 1.26 *An edge belongs to a cutset (circuit) of a graph iff it does not form a self-circuit (self-cutset) of the graph.*

Proof Suppose an edge e belongs to a cutset S of a graph G and at the same time forms a self-circuit C_s of G. Then $|C_s \cap S| = 1$ which contradicts the Orthogonality Theorem (Proposition 1.20). Thus if e belongs to a cutset then it does not form a self-circuit.

Conversely, suppose that e does not belong to a cutset. Then e cannot be incident to a pair of distinct vertices (because otherwise any vertex subset that contains one of the end vertices and does not contain the other one defines a cut containing e). Consequently e must be incident to a single vertex and hence e forms a self-loop. In a completely dual manner it can be proved that e belongs to a circuit iff it does form a self-cutset. □

The next proposition, which is usually called the Painting Theorem, is a nontrivial consequence of Propositions 1.26 and 1.20. This theorem and the Orthogonality Theorem will be the most frequently employed results within this book.

Proposition 1.27 [Painting Theorem] *Given a graph G let $\{e\}$, E_1 and E_2 form a partition of the edge set E of G, where e is an edge which does not form either a self-circuit or a self-cutset of G. Then either e forms a circuit with edges in E_1 only or a cutset with edges in E_2 only, but not both.*

Proof Let G' be the graph obtained from G by contracting all edges in E_1. Suppose that e forms a self-circuit of G'. Then e forms a circuit of G with edges in E_1 only. According to the Orthogonality Theorem (that is, Proposition 1.20), e does not form a cutset of G with edges in E_2 only, because E_1 and E_2 are disjoint. Suppose now that e does not form a self-circuit of G'. Then, according to Proposition 1.26, e belongs to a cutset of G', that is, e forms a cutset of G with edges in E_2 only. Hence, according to the Orthogonality Theorem, e does not form a circuit with edges in E_1 only, because E_1 and E_2 are disjoint. Thus, e either forms a circuit with edges in E_1 only or a cutset with edges in E_2 only, but not both. □

It is easy to see that the following seemingly stronger version of the Painting Theorem (Proposition 1.27) is also true:

Proposition 1.28 *Given a graph G let $\{E_1, \{e\}, E_2\}$ be a partition of its edge set E. Then either (a) there is a circuit of G containing e and none of the elements of E_2 or (b) there is a cutset of G containing e and none of the elements of E_1.*

Notice that in Proposition 1.28 e could be either a self-circuit or a self-cutset. The next corollary is an immediate consequence of the Painting Theorem.

Corollary 1.29 [to Proposition 1.27 (the Painting Theorem)] *Let E be the edge set of a graph G and let A be a maximal circuit-less subset of E. Let $e' \in E \backslash A$ form a circuit C'_e with edges in A only and let e belong to $(A \cap C'_e) \backslash \{e'\}$. Then e' belongs to the cutset S_e that e forms with edges of $E \backslash A$ only.*

Proof Because A is circuit-less, e does not form a circuit with edges in A only and hence, according to the Painting Theorem applied to the triple $(A \backslash \{e\}, \{e\}, E \backslash A)$, the edge e forms a cutset S_e with the edges in $E \backslash A$ only. Suppose that S_e does not contain e'. Then, $C \cap S_e = \{e\}$ which contradicts the Orthogonality Theorem. □

The following dual statement is also true:

Dual of Corollary 1.29: *Let E be the edge set of a graph G and let A be a maximal cutset-less subset of E. Let $e' \in E \backslash A$ form a cutset S'_e with edges in A only and let e belong to $A \cap S'_e$. Then e' belongs to the circuit C_e that e forms with edges of $E \backslash A$ only.*

Until now, the ring sum operation has been applied as an internal operation in the collection of all circs $\mathbf{C}(G)$ and independently in the collection of all cuts $\mathbf{S}(G)$. The next assertion, the proof of which is presented in chapter 2 (Proposition 2.33), shows a deep connection between $\mathbf{C}(G)$ and $\mathbf{S}(G)$ through the ring sum operation.

Assertion 1.30 *Given a graph G, any proper subset of its edge set E can be expressed as the ring sum of a circ and a cut of G iff the intersection of $\mathbf{C}$ and $\mathbf{S}$ is empty.*

According to Assertion 1.30 a subset of the edge set E of a graph G can be expressed as a ring sum of a circ and a cut of G iff there is not a cut of G which is at the same time a circ of G. For example, for the graph of Figure 1.1 with edge set $E = \{e_1, \ldots, e_{10}\}$ the subset $\{e_3\}$ is the ring sum of the cut $\{e_2, e_3, e_4, e_6, e_7\}$ and the circ $\{e_2, e_4, e_6, e_7\}$. An example of a graph G for which $\mathbf{C}(G)$ and $\mathbf{S}(G)$ are not disjoint is the graph of Figure 1.7. Because this graph is both bipartite and Eulerian, the edge set $E = \{e_1, \ldots, e_6\}$ forms both a circ and a cut at the same time. For this graph, there are subsets of edges which cannot be represented as the sum of a circ and a cut. Consider for example the edge subset $\{e_1\}$. According to Remark 1.7, every cut and every circ of the graph are of even cardinality. On the other hand, from Proposition 1.20, the intersection of a cut and a circ is also even. Therefore, the ring sum must be of even cardinality, which means that a single element edge subset cannot be represented as the sum of a circ and a cut.

However, if as an edge subset of a graph G we take the whole edge set E of G, then Assertion 1.30 always holds. That is, there is a cut S of G and a circ C of G such that $S \oplus C = E$. Moreover, because every edge in E must belong to the ring sum, no edge is in the intersection $C \cap S$. Therefore the ring sum $S \oplus C$ coincides with the union $S \cup C$ of disjoint edge subsets S and C, that is, the following assertion which is proved in chapter 2 (Proposition 2.34), holds.

Assertion 1.31 *For every graph G, there is partition (E', E'') of the edge set E such that E' is a cut of G and E'' is a circ of G.*

As an illustration, consider the graph in Figure 1.8. For this graph, there are four pairs of edge subsets representing cut–circ partitions. These are $(\{a, c, e\}, \{b, d, f\})$, $(\{a, b, f\}, \{c, d, e\})$, $(\{d, e, f\}, \{a, b, c\})$, and $(\{b, c, d\}, \{a, e, f\})$. Notice that for this graph, there are edge subsets that are at the same time circs and cuts. These are $\{a, b, d, e\}$, $\{b, c, e, f\}$ and $\{a, c, d, f\}$.

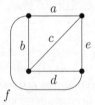

Figure 1.8. A graph illustrating cut–circ partitions.

1.5 Edge-separators and connectivity

In section 1.1 we introduced vertex-separators of a graph. Here we introduce two types of edge-separators which, unlike vertex-separators, ignore isolated vertices.

A subset E_1 of the edge set E of a graph G is said to be a *circuit-separator* if each circuit of G is contained either in E_1 or in $E \backslash E_1$. It is easy to see that the union and intersection of circuit-separators is a circuit-separator and the complement of a circuit-separator is also a circuit-separator. In particular, the empty set, every one element set consisting of self-circuit edge or self-cutset edge and the whole edge set E are circuit-separators of G. A circuit-separator is an *elementary circuit-separator* if it contains no other circuit-separator of G as a proper subset. Notice that both a self-circuit edge and a self-cutset edge are elementary circuit-separators.

A graph G is said to be *circuit-connected* if it has no circuit-separators other then the empty set and the set E. Let $\{E_1, \ldots, E_q\}$ be the set of all elementary circuit-separators of a graph G and let G_j be the edge-induced subgraph associated with an elementary circuit-separator E_j. Obviously each G_j itself is circuit-connected. We say that G_j is a *circuit-connected component* of G.

A subset E_1 of the edge set E of a graph G is said to be a *cutset-separator* of G if each cutset of G is contained either in E_1 or in $E \backslash E_1$. It is easy to see that the union and intersection of cutset-separators is a cutset-separator and the complement of a cutset-separator is also a cutset-separator. In particular, the empty set, every one element set consisting of self-circuit edge or self-cutset edge and the whole edge set E are cutset-separators. A cutset-separator is an *elementary cutset-separator* if it contains no other cutset-separator of G as a proper subset. Notice that both self-circuit edge and self-cutset edge are elementary cutset-separators.

Let G be a graph with edge set E. Then we say that G is *cutset-connected* if it has no cutset-separators other then the empty set and the set E. Let $\{E_1, \ldots, E_q\}$ be the set of all elementary cutset-separators of a graph G and let G_k be the edge-induced subgraph associated with an elementary cutset-separator E_k. Obviously each G_k itself is cutset-connected. We say that G_k is a *cutset-connected component* of G.

The following proposition holds.

Proposition 1.32 *Let G be a graph with the edge set E and let E' and $E'' \subset E$ be two circuit- (cutset-) separators of G. Then $E' \backslash E''$ is also a circuit- (cutset-) separator of G.*

Proof Since $E'' \subset E'$ and E'' is a circuit (cutset)-separator it follows that no circuit (cutset) contains edges from both E' and E''. But E' is also a circuit (cutset)-separator and hence each circuit (cutset) of G belongs precisely to one of $E' \backslash E''$ or E'' or $E \backslash E'$. Consequently each circuit (cutset) of G either belongs to $E' \backslash E''$ or belongs to $E \backslash (E' \backslash E'')$. $\square$

Corollary 1.33 [to Proposition 1.32] *Any circuit- (cutset-) separator of a graph G is either an elementary circuit- (cutset-) separator of G or a union of disjoint elementary circuit- (cutset-) separators of G.*

Thus, the elementary circuit- (cutset-) separators of G are disjoint nonempty subsets of E whose union is E. The next proposition shows that circuit and cutset connectivity are two equivalent concepts.

Proposition 1.34 *A subset of edges of a graph is a cutset-separator iff it is a circuit-separator.*

Proof Clearly, every self-circuit and every self-cutset is both an elementary cutset-separator and an elementary circuit-separator. Suppose A (which is neither a self-circuit nor a self-cutset) is a cutset-separator but not a circuit-separator. Then there is a circuit C of a graph G containing two regular edges a and b such that a belongs to A and b belongs to $E\backslash A$. Let G' be the graph obtained from G by contracting all edges of G that belong to $C\backslash\{a,b\}$. From Lemma 1.9 $\{a,b\}$ is a circuit of G'. Therefore, a and b do not form self-circuits and so (Proposition 1.26) each of them belongs to a cutset. Also, a and b are parallel edges in G' and therefore a belongs to every cutset that b belongs to and vice versa. Let S be such a cutset. According to Lemma 1.8 any cutset of G' is a cutset of G and hence S is a cutset of G. But S obviously does not belong entirely either to A or to $E\backslash A$ which contradicts the assumption that A is a cutset-separator.

Conversely, suppose A (which is neither a self-circuit nor a self-cutset) is a circuit-separator but not a cutset-separator. Then there is a cutset S of a graph G containing two regular edges a and b such that a belongs to A and b belongs to $E\backslash A$. Let G'' be the graph obtained form G by removing all edges of G that belong to $S\backslash\{a,b\}$. From Lemma 1.9 $\{a,b\}$ is a cutset of G''. Therefore, a and b do not form self-cutsets and so (Proposition 1.26) each of them belongs to a circuit. On the other hand, $\{a,b\}$ is a cutset of G'' and therefore a must belong to every circuit that contains b. Let C be such a circuit. From Lemma 1.8 any circuit of G'' is a circuit of G and hence C is a circuit of G. Thus obviously C does not belong entirely either to A or to $E\backslash A$ which contradicts the assumption that A is a circuit-separator.   □

A graph may be circuit-connected (respectively, cutset-connected) even though it has isolated vertices. This is because circuit-connectness (cutset-connectness) is an edge-oriented notion which ignores isolated vertices.

Assertion 1.35 *A graph G without isolated vertices is 2-connected iff it is circuit-connected.*

As an immediate consequence of Proposition 1.34 and Assertion 1.35, the following statement is also true:

Dual of Assertion 1.35: *A graph G without isolated vertices is 2-connected iff it is cutset-connected.*

Let G be a 1-connected graph with edge set E and let $\{E_1,\ldots,E_q\}$ be the set of all elementary cutset-separators of a graph G. Let G_k be the edge-induced subgraph associated with an elementary cutset-separator E_k. Obviously each G_k itself is cutset-connected. Then we say that G_k is a *2-connected component* of G.

1.6   Equivalence relations among graphs

The answer to the question of whether two graphs with the same number of vertices and the same number of edges can be treated as equal, depends upon the features that are pertinent. Informally, two graphs, G_1 and G_2 can be treated as equal or *isomorphic* if there are two bijections, one between vertices of G_1 and G_2, and another between edges of G_1 and G_2 preserving incidences between vertices and edges. The binary relation 'to be isomorphic' is an equivalence relation in the set of all graphs so we may speak about classes of isomorphism. The fact that two graphs G_1 and G_2 are isomorphic is denoted by $G_1 = G_2$. Graph theory is concerned only with those properties of graphs that are invariant under isomorphism.

Apart from isomorphism, there are other equivalence relations pertinent to the question of whether two graphs are the same. We introduce two such equivalence relations: *1-isomorphism* and *2-isomorphism*.

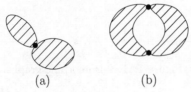

(a)   (b)

Figure 1.9.   Graph splitting structures: (a) 1-splitting structure; (b) 2-splitting structure.

Let G be a graph and let E be its edge set. Assume that there is a partition (E', E'') of E such that the associated edge-induced subgraphs G' and G'' have exactly one vertex v_o in common. A general structure of such a graph is shown in Figure 1.9(a). It is clear that such a vertex v exists for a graph G iff the associated star is a separable star of G. Consider the operation which consists of splitting v into two copies v' and v'' so that v' belongs to the vertex set of G' and v'' belongs to the vertex set of G''. Call this operation the *1-splitting operation*. We say that two graphs are *1-isomorphic* if they are isomorphic after a number of splitting operations. Consider for example the two graphs of Figure 1.10. If we split vertex $\check{v}$ of the first graph and vertex $\hat{v}$ of the second one then we obtain isomorphic graphs, each with two 2-connected components. Consequently, these graphs are 1-isomorphic.

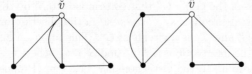

Figure 1.10. 1-isomorphic graphs.

Remark 1.36 *Two graphs G' and G'', are 1-isomorphic iff the corresponding 2-connected components are isomorphic.*

Suppose that for a graph G there is a partition (E', E'') of the edge set E such that the associated edge-induced subgraphs G' and G'' have exactly two vertices, say v_1 and v_2, in common. A general structure of such a graph is presented in Figure 1.9(b). Consider the operation which consists of splitting v_1 into two copies v_1' and v_1'' and v_2 into two copies v_2' and v_2'' such that v_1' and v_2' belong to the vertex set of G' and v_1'' and v_2'' belong to the vertex set of G''. Call this operation: the *2-splitting operation*. Finally, identify v_1' and v_2' and identify v_1'' and v_2''. The overall operation is called a *twisting operation* at v_1 and v_2. Two graphs are *2-isomorphic* if they are isomorphic after a number of 2-splitting and/or twisting operations.

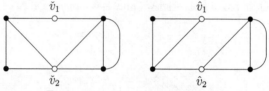

Figure 1.11. 2-isomorphic graphs.

For example, Figure 1.11 shows two graphs which are 2-isomorphic. If, we perform consecutively a 2-splitting and a twisting operation on the pair of vertices $(\check{v}_1, \check{v}_2)$ of the first graph then we obtain a graph which is isomorphic to the second. Alternatively, performing the same two operations on the pair $(\hat{v}_1, \hat{v}_2)$ of the second graph produces a graph which is isomorphic to the first.

The next assertion is a well-known result, which we present without proof.

Assertion 1.37 *Two graphs are 2-isomorphic iff there exists a bijection between their edge sets which preserves circuits.*

Proposition 1.38 *Any bijection between the edge sets of two graphs that preserves circuits, also preserves cutsets and vice versa.*

Proof Let G' and G'' be two graphs for which there exists a bijection between their edge sets that preserves circuits. Let S' be a cutset of G' and let S'' be the set of edges of the graph G'' that corresponds to S' by this bijection. Let C' and C'' be two circuits that correspond in the bijection providing that C' is a circuit of G' and C'' is a circuit of G''. This bijection also preserves the cardinality of the intersections $S' \cap C'$ and $S'' \cap C''$ where C' is any circuit of G'. Since, according to the Orthogonality Theorem (Proposition 1.20), the cardinality of $S' \cap C'$ is even, it follows that the cardinality of $S'' \cap C''$ is also even. Then, from Proposition 1.23 S'' is a cut. Suppose now that S'' contains a cutset S_1'' as a proper subset. Interchanging the places of graphs G' and G'' and using the same argument as in previous consideration we conclude that S' contains a cutset S_1' as a proper subset, which is a contradiction. Therefore, S'' is a cutset, and hence the bijection which preserves circuits also preserves cutsets. Using dual arguments we can prove that the bijection which preserves cutsets also preserves circuits, which completes the proof.

$\square$

As an immediate consequence of Assertion 1.37 and Proposition 1.38, the following dual assertion also holds:

Dual of Assertion 1.37 *Two graphs are 2-isomorphic iff there exists a bijection between their edge sets which preserves cutsets.*

For example, for the pair of graphs of Figure 1.11 the isomorphism described in Figure 1.12, preserves both circuits and cutsets.

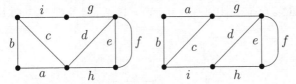

Figure 1.12. 2-isomorphic graphs with labels.

These two graphs are 2-isomorphic and have the same triple $(E, \mathbf{C}, \mathbf{S})$ where:

$$\mathbf{S} = \{\{a,b\}, \{i,g\}, \{b,c,i\}, \{b,c,g\}, \{e,f,h\}, \{d,h,i\}, \{a,c,i\},$$
$$\{a, c, g\}, \{d, g, h\}, \{a, c, d, h\}, \{d, e, f, g\},$$
$$\{d, e, f, i\}, \{a, c, d, h\}, \{b, c, d, h\}, \{a, c, d, e, f\}, \{b, c, d, e, f\}\}$$

and

$$\mathbf{C} = \{\{e, f\}, \{a, b, c\}, \{d, e, h\}, \{d, f, h\}, \{c, d, g, i\},$$
$$\{a, b, d, g, i\}, \{a, b, e, g, i, h\}, \{a, b, f, g, i, h\}, \{c, e, g, i, h\}, \{c, f, g, i, h\}\}.$$

The binary relation 'to be 2-isomorphic' is obviously an equivalence relation in the set of all graphs. Among the classes of 2-isomorphic graphs we can define the concept of duality. Two classes $\mathbf{G}'$ and $\mathbf{G}''$ of 2-isomorphic graphs are dual to each other if for every pair of graphs, (G', G'') such that $G' \in \mathbf{G}'$ and $G'' \in \mathbf{G}''$ there is a bijection between the edge sets of these graphs so that the circuits of one of these graphs are cutsets of another, and vice versa. Denote by $\mathbf{G}^*$ the class of 2-isomorphic graphs dual to the class $\mathbf{G}$. Then $\mathbf{G}^{**} = \mathbf{G}$ holds.

Therefore, with each equivalence class of 2-isomorphic graphs with edge set E we can associate the unique triple $(E, \mathbf{C}, \mathbf{S})$ where $\mathbf{C}$ is the collection of circuits and $\mathbf{S}$ is the collection of cutsets. This triple defines a concept called a *graphoid* (see section 6 of chapter 2 for the definition of graphoid and more details). Thus the concept of 2-isomorphism naturally leads to the domain of graphoids and a vertex-independent view of graphs.

1.7 Directed graphs

A graph is directed (is a *digraph*) if a direction is assigned to each of its edges. It is a partially directed graph if some of its edges are directed and some are not. For each directed or partially directed graph there is exactly one graph associated with it. The converse is not true; several digraphs may be associated with a particular nonoriented graph.

For example a digraph and the associated graph are presented in Figures 1.13 and 1.14 respectively. For both graphs $E = \{a, b, c, d, e, f, g, h, k\}$ is the set of all edges (oriented or not) and $V = \{1, 2, 3, 4, 5, 6\}$ is the set of vertices.

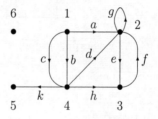

Figure 1.13. A digraph.

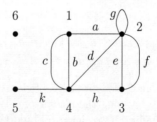

Figure 1.14. The graph associated with the digraph of Figure 1.13.

Clearly, every directed edge leaves exactly one vertex (its out-vertex) and enters exactly one vertex (its in-vertex). This is described by an in-map f' and an out-map f''.

Formally, a *directed graph* or *digraph* is a quadruple (E, V, f', f''), where E and $V \neq \emptyset$ are disjoint finite sets and f', f'' are maps from E to V. For example for the digraph of Figure 1.13 $f'(a) = f'(d) = f'(f) = f'(g) = \{2\}, f'(e) = f'(h) = \{3\}, f'(b) = f'(c) = \{4\}, f'(k) = \{5\}; f''(a) = f''(b) = f''(c) = \{1\}, f''(e) = f''(g) = \{2\}, f''(f) = \{3\}, f''(d) = f''(k) = f''(h) = \{4\}$.

The maps f' and f'' are not generally surjective. That is, $f'(E), f''(E) \subseteq V$. The elements of V are called vertices and the elements of E are called *directed edges*. The maps $f'^{*}: V \rightarrow 2^E$, given by $f'^{*}(v) = \{e \in E | v \in f'(e)\}$ and $f''^{*}: V \rightarrow 2^E$, given by $f''^{*}(v) = \{e \in E | v \in f''(e)\}$ are called the transposes of f' and f'', respectively. Since $v \in f'(e)$ iff $e \in f'^{*}(v)$ and since $v \in f''(e)$ iff $e \in f''^{*}(v)$, we have $f'^{**}(e) = \{v \in V \mid e \in f'^{*}(v)\} = f'(e)$ and $f''^{**}(e) = \{v \in V \mid e \in f''^{*}(v)\} = f''(e)$. If $e \in f'^{*}(v)$ we say that e is incident *to* v. If $e \in f''^{*}(v)$ we say that e is incident *from* v. The positive integers $|f'^{*}(v)|$ and $|f''^{*}(v)|$ are called the *in-degree* (denoted by $d'(v)$) and the *out-degree* (denoted by $d''(v)$) of v. If $f'^{*}(v) = \emptyset$ and $f''^{*}(v) = \emptyset$ the vertex v is an *isolated vertex*. If $f'(e_1) = f'(e_2)$ and $f''(e_1) = f''(e_2)$ then we say e_1 and e_2 are *parallel directed edges*. A digraph without parallel edges and self-loops is a *simple digraph*. The digraphs (E_1, V_1, f_1', f_1'') and (E_2, V_2, f_2', f_2'') are *isomorphic digraphs* if there are bijections $p: V_1 \rightarrow V_2$ and $q: E_1 \rightarrow E_2$ such that $f_2' = p \circ f_1' \circ q^{-1}$ and $f_2'' = p \circ f_1'' \circ q^{-1}$ (see Figure 1.15).

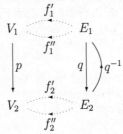

Figure 1.15. Isomorphism diagram.

Let (E, f', f'', V) be a digraph. Then the unique graph associated with the digraph is the triple (E, f, V), where $f: E \rightarrow 2^V$ is a map given by $f(e) = \{f'(e), f''(e)\}$, for all $e \in E$. For the graph in Figure 1.14,

$$
\begin{aligned}
f(a) &= \{f'(a), f''(a)\} = \{1, 2\}, \\
f(b) &= \{f'(b), f''(b)\} = \{1, 4\}, \\
f(c) &= \{f'(c), f''(c)\} = \{1, 4\}, \\
f(d) &= \{f'(d), f''(d)\} = \{2, 4\}, \\
f(e) &= \{f'(e), f''(e)\} = \{2, 3\}, \\
f(f) &= \{f'(f), f''(f)\} = \{2, 3\}, \\
f(g) &= \{f'(g), f''(g)\} = \{2\}, \\
\end{aligned}
$$

$$f(h) = \{f'(h), f''(h)\} = \{3, 4\},$$
$$f(k) = \{f'(k), f''(k)\} = \{4, 5\}.$$

Circuits of digraphs are the same as circuits of the associated graph. In the edge set of each circuit we can introduce a cyclic order such that every two consecutive edges in this order are adjacent (share a common vertex). It is easy to see that there are exactly two such orders for each circuit of a graph and that two edges are adjacent in one of these orders iff they are adjacent in another one. Thus, for every circuit we can introduce a circuit orientation by choosing one of two possible cyclic orders. We shall call them dual cyclic orders.

Cutsets of digraphs are the same as cutsets of the associated graph. Because each cutset of a graph is associated with a nontrivial partition $\{V_1, V_2\}$ of the vertex set V of the graph, we can introduce a cutset orientation by choosing one of two possible orders (V_1, V_2) or (V_2, V_1) of the unordered set $\{V_1, V_2\}$. We shall call them dual cutset orders.

A circuit (cutset) of a digraph is a *uniform circuit* (*uniform cutset*) if all the directed edges of the circuit (cutset) have the same orientation within the circuit (cutset). A graph is said to be a *partially directed graph* if some of its edges are directed and some are not. We shall denote the set of all its directed edges by P_o and the set of all its nondirected edges by P. Given a partially directed graph with the edge set $P_o \cup P$, let (P', P'') be an arbitrary partition of the set P. Then, the following statement, called the *Painting Theorem for partially directed graphs* holds:

Assertion 1.39 *Each directed edge of a partially directed graph either belongs to a uniform circuit made of edges in P_o and P' only or belongs to a uniform cutset made of edges in P_o and P'' only.*

Note that in this theorem nothing has been assumed either regarding the way in which the directed edges are directed or how nondirected edges are divided into the subsets P' and P''. In the usual formulation of the Painting Theorem, for the sake of mnemonics, the colours green, red and blue are used to denote the members of the sets P_o, P' and P'', respectively. This explains the name of the Theorem.

As an illustration, consider the partially directed graph of Figure 1.16 where every edge belongs to exactly one of the following sets $P_o = \{b, d, e, j, h\}$, $P' = \{a, f, g\}$ or $P'' = \{i, c\}$. The colours green (dotted lines), red (bold lines) and blue (plain lines) are used to denote the members of the sets P_o, P' and P'', respectively. By inspection, $e \in P_o$ forms a uniform circuit $\{e, f, g\}$ with edges in $P_o \cup P'$ only, but not a cutset with edges in $P_o \cup P''$ only. On the other hand, b forms a uniform cutset with edges in $P_o \cup P''$ only, but not a uniform circuit with edges in $P_o \cup P'$ only.

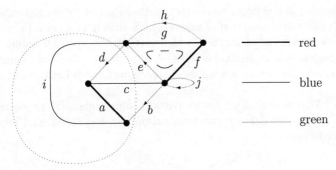

Figure 1.16. A partially oriented graph.

1.8   Networks and multiports

Consider a collection of abstract objects each with two access points, called terminals. At this stage the objects are not restricted; for example they may be electrical, mechanical, thermodynamic, depending on what physical attributes we associate with them. In order to obtain a mathematical model, with each pair of terminals we associate two types of oriented scalar variables: 'through' variables and 'across' variables. Such pairs of variables are called conjugate variables since their scalar product always has the dimension of power. Typical conjugate variables are current and voltage (in the case of electrical objects), force and velocity (in the case of mechanical objects), temperature and entropy change (in the case of thermodynamic objects). It is generally impossible to specify in advance the actual orientations of these variables. We therefore set up a so-called *frame of reference* in terms of which the actual orientation of current and voltage vectors can be specified. For each two-terminal object, a reference frame merely consists of arbitrarily assigning a reference orientation for the through variable by an arrow (oriented from one terminal to another) and a reference orientation for the across variable, by associating the sign (+) with one terminal and the sign (−) with another. Thus the order of the terminals of a two-terminal object defines the orientation of the associated conjugate scalar variables.

A (lumped) *network* consists of a set of two-terminal objects that are united by two forms of interdependencies:

- By *joining object terminals* at disjoint junctions (*galvanic connection*).

- By *coupling* the objects through a continuous media (*non-galvanic connection*).

The collection of two-terminal objects of a network is always *assumed to be closed* in the sense that there is no object external to the collection influencing objects in it through galvanic connection or through coupling.

Any junction in a network where terminals are joined together is called a node. Galvanic interconnection of two-terminal objects of a network can be described by an associated graph in which vertices correspond to nodes and edges correspond to two-terminal objects. A two-terminal object is commonly referred to as a port. From now on we shall say port instead of two-terminal object.

Figure 1.17 schematically shows a network which might be an electrical circuit. Eleven boxes indicate two-terminal objects and the shaded regions illustrate couplings.

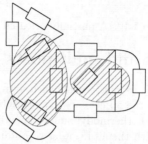

Figure 1.17. A network structure that describes the interdependance of two-terminal objects.

If the reference orientation for the through variable is from + to − then we say that the orientations of the conjugate pair of variables are in accordance. In this case, to introduce reference orientations for a port it is enough to assign the orientation for only one of the variables of a conjugate pair. A network for which all ports are directed is called an oriented network.

In the set of ports of a network consider a partition such that no two ports from different parts of the partition are coupled. In other words every member of the partition is closed with respect to couplings. Clearly, there could be several such partitions but there is only one with the property: no member of the partition contains a member of some other partition as a proper subset. Such a partition always exists and is unique. We call it a maximum cardinality partition of the network ports by means of couplings. Every member of such a partition of network ports is called a *multiport* of the network. In particular if it contains n ports it is called an *n-port*. We can regard multiports as basic network units. From this point of view, every network is made of a number of multiports that are interconnected strictly by means of galvanic connections. In particular, a maximum cardinality partition by means of couplings may consist of only one member which coincides with the collection of all network ports.

Similarly, we may define the maximum cardinality partition of the ports of a network such that every member is closed with respect to galvanic connections. Again, such a partition exists and is unique. Every member of such partition of network ports is called a *galvanic component* of the network. It is easy to see that a collection of network ports forms a galvanic object iff the corresponding graph induced by this collection is 1-connected. In particular, a maximum cardinality partition of the network ports with respect to galvanic connections may consist of only one member coincident with the set of all network ports.

Suppose that each port of a network has an associated pair of well defined oriented scalars: a *current* passing through the port from one terminal to another and a *voltage* between the two terminals. Then, the network is called an *electrical network*. Two-terminal electrical objects could then be resistors, capacitors, coils and so on. A wire between two terminals provides a galvanic connection, whereas an electromagnetic field between two coils introduces couplings, that is, a non-galvanic connection. If the reference orientation for the through variable is from $+$ to $-$ then we say that the orientations of the conjugate pair of variables are in accordance, as illustrated in Figure 1.18. In this case, to introduce reference orientations for a port it is enough to assign the orientation for only one of the variables of a conjugate pair. A network for which all ports are directed is called an *oriented electrical network*.

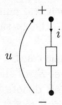

Figure 1.18. Reference port orientations.

A map whose domain is time is called a *signal*. Normally, the time domain is simply the set of real numbers which we denote by R. A signal whose codomain is also the set of real numbers we call a real signal.

Generally speaking, an electrical n-port is characterized by a set $\mathcal{E}_n$ of all possible $2n$-tuples of real valued signals $(u_1(t), i_1(t); \ldots ; u_n(t), i_n(t))$ that are simultaneously allowed on the ports, providing that $u_k(t)$ is the voltage signal and $i_k(t)$ is the current signal associated with a port k, for $k = 1, \ldots, n$. We assume that all these signals belong to the same universal set of real-valued signals which we shall denote by S. Clearly, $\mathcal{E}_n \subseteq S^{2n}$ and therefore $\mathcal{E}_n$ defines a relation in the signal set S. This explains why the set $\mathcal{E}_n$ of all allowed $2n$-tuples of an n-port is called a *constitutive relation*. A general n-port can be also viewed as an algorithm which decides which $2n$-tuples of signals are allowed and which are not.

For $n \geq 1$, consider an internal binary additive operation $+$ defined on S^{2n} and an external multiplicative operation $(\circ)$ defined from R to S^{2n}, induced by analogous operation in the real field $(\mathsf{R}, +, \cdot)$, as follows:

$$(u_1'(t), i_1'(t); \ldots; u_n'(t), i_n'(t)) + (u_1''(t), i_1''(t); \ldots; u_n''(t), i_n''(t))$$
$$\stackrel{\text{def}}{=} (u_1'(t) + u_1''(t), i_1'(t) + i_1''(t); \ldots; u_n'(t) + u_n''(t), i_n'(t) + i_n''(t)),$$

$$\alpha \circ (u_1(t), i_1(t); \ldots; u_n(t), i_n(t)) \stackrel{\text{def}}{=} (\alpha u_1(t), \alpha i_1(t); \ldots; \alpha u_n(t), \alpha i_n(t)).$$

It is easy to see that $(S^{2n}, +, \circ)$ is a *linear* (vector) space over R. Let $\mathcal{E}_n \subseteq S^{2n}$ be the constitutive relation of an n-port. If $(\mathcal{E}_n, +, \circ)$ is linear (vector) subspace of S^{2n}, than we say that the n-port characterized by $\mathcal{E}_n$ is a *linear n-port*. Let $(u_{1o}(t), i_{1o}(t); \ldots; u_{no}(t), i_{no}(t))$ be an arbitrary element in $\mathcal{E}_n$. Denote

$$\mathcal{E}_n^o = \mathcal{E}_n \setminus \{(u_{1o}(t), i_{1o}(t); \ldots; u_{no}(t), i_{no}(t)\}.$$

If for every $2n$-tuple $(u_{1o}(t), i_{1o}(t); \ldots; u_{no}(t), i_{no}(t)) \in \mathcal{E}_n$, the triple $(\mathcal{E}_n^o, +, \circ)$ is a linear subspace of S^{2n}, then $\mathcal{E}_n$ is said to be an *affine* space over R. Then we say that the n-port characterized by $\mathcal{E}_n$ is an *affine n-port*. If $\mathcal{E}_n$ is an affine n-port then $\mathcal{E}_n^o$ is the unique linear n-port associated with the affine one. We call it the unique linear n-port *parallel* to the affine n-port. An n-port which is neither linear nor affine is called a *nonlinear n-port*. We say that an n-port is *time-invariant* if for each real number T and for each $2n$-tuple $(u_1(t), i_1(t); \ldots; u_n(t), i_n(t))$ from $\mathcal{E}_n$, the $2n$-tuple $(u_1(t-T), i_1(t-T); \ldots; u_n(t-T), i_n(t-T))$ is also in $\mathcal{E}_n$.

The set S^{2n} of all $2n$-tuples signals can be viewed as the set of *all* maps from R to R^{2n}. Therefore, any subset $\mathcal{E}_n$ of the set S^{2n} is as a collection of maps from R to R^{2n}. For some $t \in \mathsf{R}$ denote by E_n^t the set obtained from $\mathcal{E}_n$ by restricting all maps in $\mathcal{E}_n$ from R to $\{t\}$. We call E_n^t the *chart* of $\mathcal{E}_n$ at the moment $t \in \mathsf{R}$. At any moment $t \in \mathsf{R}$ the associated chart E_n^t is the set of all $2n$-tuples of *values* of voltages and currents that are allowed by the n-port at the same moment $t \in \mathsf{R}$. The family $\{\, E_n^t \mid E_n^t \subseteq \mathsf{R}^{2n}, t \in \mathsf{R}\}$ of all charts of $\mathcal{E}_n$ we call the *atlas* of $\mathcal{E}_n$. Notice that with every set $\mathcal{E}_n$ there is a uniquely associated atlas. Let $\mathcal{S}_n$ be the set of *all* maps from R to R^{2n} that are in accordance with the atlas of $\mathcal{E}_n$, in the following sense: for every $t \in \mathsf{R}$ the image of t belongs to $E_n^t \subseteq \mathsf{R}^{2n}$. Clearly, $\mathcal{E}_n \subseteq \mathcal{S}_n \subseteq S^{2n}$. If $\mathcal{E}_n = \mathcal{S}_n$ then we say that an n-port is a *resistive n-port*. Therefore, a resistive n-port is completely described by its atlas. A general resistive n-port could be seen as an algorithm which decides at every moment $t \in \mathsf{R}$ which $2n$-tuples from E_n^t are allowed and which are not. When the n-port is time-invariant all charts coincide (the atlas consists of only one chart), that is, there is an $E_n \subseteq \mathsf{R}^{2n}$ such that for all $t \in \mathsf{R}$, $E_n^t = E_n$. Consequently, a resistive n-port which is time-invariant is uniquely determined by a set $E_n \subseteq \mathsf{R}^{2n}$ of $2n$-tuples of values of voltages and currents.

Usually we deal with resistive n-ports whose atlas can be interpreted as the set of solutions of a system of algebraic time-varying equations in terms of port voltages and currents. In this interpretation the set of solutions of the system with fixed parameter $t \in \mathsf{R}$ coincides with the chart E_n^t of the n-port. The equations of the system we call constitutive relations of a resistive n-port. We say that a resistive n-port is a *regular n-port* if it is characterized by n algebraic equations. Otherwise, if the number of ports and the number of algebraic equations differ, the n-port is a *singular n-port*.

For example, a regular time-invariant resistive 1-port is characterized by an algebraic equation of the form $f(u, i) = 0$ where the pair (u, i) is a pair of port variables. A linear version of a regular, time-invariant, 1-port resistor is characterized by an equation of the type $au + bi = 0$, where $a^2 + b^2 \neq 0$. If both $a, b \neq 0$, then this regular 1-port resistor is called an *Ohm's resistor*. Otherwise if $a = 0$ and $b \neq 0$, it is an *open-circuit*; if $a \neq 0$ and $b = 0$, it is a *short-circuit*. An example of a singular resistive 1-port is the so-called *nullator*, which is characterized by two relations $u = 0$ and $i = 0$. Another type of singular resistive 1-port is the so called *norator* for which the number of algebraic equations is zero, that is, every possible pair (u, i) is allowed. A resistive n-port is linear (respectively, affine, nonlinear) iff the corresponding algebraic equations are linear (respectively, affine, nonlinear). An *independent voltage source* $u = u_g(t)$ and an *independent current source* $i = i_g(t)$ are examples of affine, time-variant 1-port resistors. Note that the independent voltage source for which $u_g(t) = 0$ coincides with a short-circuit and the independent current source for which $i_g(t) = 0$ coincides with an open-circuit. Some linear and affine 1-port resistors are presented in Figure 1.19.

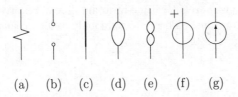

(a)	(b)	(c)	(d)	(e)	(f)	(g)

Figure 1.19. Some linear resistive 1-ports: (a) Ohm's resistor; (b) open-circuit; (c) short-circuit; (d) nullator; (e) norator; (f) independent voltage source; (g) independent current source.

We say that a regular, resistive, time-invariant 1-port is u-controlled if it is characterized by a relation of the type $u = \hat{u}(i)$. It is i-controlled if it is characterized by a relation of the type $i = \hat{i}(u)$. An example of a u-controlled, affine, time-variant 1-port resistor is $i = Gu + i_g(t)$ and an example of an i-controlled, nonlinear, time-invariant, 1-port resistor is $u = Ri^2$.

Similarly, a regular, resistive, time-invariant 2-port is characterized by two algebraic equations of the form: $f_1(u_1, i_1; u_2, i_2) = 0$ and $f_2(u_1, i_1; u_2, i_2) = 0$, where (u_1, i_1) and (u_2, i_2) are pairs of port variables. Some important

examples of regular, time-invariant, linear, resistive 2-ports are presented in
Figure 1.20. An *ideal transformer* is defined by the algebraic constitutive
relations $u_1 = mu_2$ and $i_2 = -mi_1$, where m is a real parameter; an *ideal
gyrator*, is defined by the algebraic constitutive relations $u_1 = ri_2$ and $u_2 =
-ri_1$, where r is a real parameter. A pair (nullator, norator), is called an
ideal operational amplifier or simply *nulor*. It can be considered as a time-
invariant, linear, resistive 2-port, with algebraic constitutive relations $u_1 = 0$
and $i_1 = 0$.

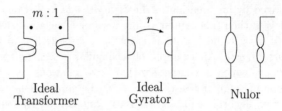

Ideal   Ideal
Transformer   Gyrator   Nulor

Figure 1.20. Some linear resistive 2-ports.

It is interesting that the nulor is obviously regular although both the
nullator and the norator are singular. This is because the nullator is characterized with two equations and norator with no equations which all together
gives two equations for two ports. From this point of view, it is natural to
treat nullator-norator pairs as nontrivial 2-ports, although there is no explicit
couplings between a nulor's ports, that is, between a nullator and a norator.

Four more types of regular, time-invariant, linear, resistive 2-ports are also
of particular importance. These are commonly called *dependent* or *controlled
sources* (see Figure 1.21).

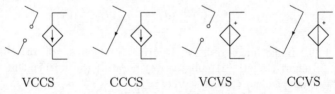

VCCS   CCCS   VCVS   CCVS

Figure 1.21. Four types of controlled sources.

Thus, the *voltage-controlled voltage source* (VCVS) is defined by the following two relations: $i_1 = 0$ and $u_2 = au_1$, where a is a real parameter without
dimension. The *current-controlled voltage source* (CCVS) is defined by the
following two relations: $u_1 = 0$ and $u_2 = ri_1$, where r is a real parameter
with a resistance dimension. The *voltage-controlled current source* (VCCS)
is defined by the following two relations: $i_1 = 0$ and $i_2 = gu_1$, where g is a
real parameter with a conductance dimension. The *current-controlled current*

source (CCCS) is defined by the following two relations: $u_1 = 0$ and $i_2 = bi_1$, where b is a real parameter without dimension.

Some of the multiports can be represented as networks made up of other multiports. For example, the following representations hold:

- every affine n-port can be represented via the associated 'parallel' linear n-port augmented at each port k with a pair of voltage and current independent sources $(u_{g_k}(t), i_{g_k}(t))$ such that the resulting $2n$-tuple $(u_{g_1}(t), i_{g_1}(t); \ldots ; u_{g_n}(t), i_{g_n}(t))$ is a $2n$-tuple allowed by the affine n-port. For example the representation of a 1-port resistor is shown in Figure 1.22.

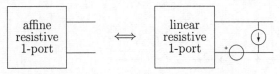

Figure 1.22. Modelling an affine 1-port via an associated linear 1-port and a pair of independent sources.

- every linear resistive n-port can be represented as a network made up of controlled sources and/or Ohm's resistors.

Accordingly, *every affine resistive n-port can be represented as a network made up of independent sources, controlled sources and Ohm's resistors.* For example a nullator and a norator can be represented via controlled sources, as shown in Figure 1.23.

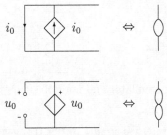

Figure 1.23. Modelling a nullator and a norator via CCCS and VCVS, respectively.

- every controlled source can be represented as a network made up of Ohm's resistors and equal numbers of nullators and norators.

Hence, *every affine resistive n-port can be represented as a network made up of independent sources, Ohm's resistors and equal numbers of nullators and norators.* For example, a gyrator can be modelled by using two Ohm's resistors (one positive and one negative), three nullators and three norators, as shown in Figure 1.24.

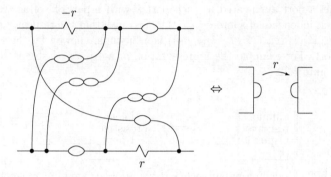

Figure 1.24. Modelling an ideal gyrator via nullator-norator pairs and Ohm's resistors.

- a nullator and a norator can each be modelled by two Ohm's resistors (one positive and one negative) and a gyrator, as shown in Figure 1.25.

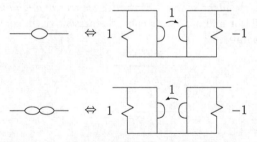

Figure 1.25. Modelling a nullator and norator via a gyrator and Ohm's resistors.

Thus, *every affine resistive n-port can be represented as a network made up of independent sources, gyrators and Ohm's resistors.* For example, every ideal transformer can be represented as a cascade connection of two gyrators, as is shown in Figure 1.26.

In conclusion, *every network made of regular affine resistive n-ports can be transformed into an equivalent network which consists of elements that belong to one of the following minimal collections of constitutive elements:*

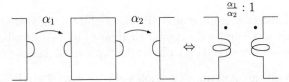

Figure 1.26. Modelling an ideal transformer via ideal gyrators.

- *Ohm's resistors, controlled sources and independent sources.*

- *Ohm's resistors, nullators, norators and independent sources.*

- *Ohm's resistors, gyrators and independent sources.*

Moreover, every negative resistive n-port can be represented as a network made up of positive Ohm's resistors and equal number of nullators and norators (see Figure 1.27).

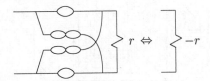

Figure 1.27. Modelling a negative Ohm's resistor via nullator and norator pairs and a positive Ohm's resistor.

Thus, in all of previous cases that describe representations of affine resistive n-ports via collections of constitutive elements we can replace the term 'Ohm's resistor' with the term '*positive* Ohm's resistors'.

1.9 Kirchhoff's laws

We can associate a unique directed graph with any oriented network, called the *network digraph*. Reference port orientations of currents in the network correspond to edge directions in the digraph. This directed graph describes the galvanic connections (the topology) of the network. We assume that the orientation of voltage and the orientation of currents at each network port are in accordance, as shown in Figure 1.18. A circuit of a directed graph is an *oriented circuit* if there is a bijection between the vertex set of the corresponding edge-induced subgraph and a circularly ordered subset of distinct integers. Two vertices are joined by an edge iff their maps according to this bijection are consecutively ordered integers. The circular order of the subset of integers induces an orientation (one of two possible) of the circuit.

A cutset of a directed graph is an *oriented cutset* if the associated 2-partition of the vertex set is ordered. Again, there are just two possibilities. The following two axiom serve as a basis for the topological analysis of networks:

Network Voltage Axiom: [Kirchhoff's voltage law] *The algebraic sum of voltage signals associated with the edges that belong to an oriented circuit of a network, is equal to zero. By algebraic, we mean that every voltage appears in this sum with a plus sign if the orientation of the corresponding edge and the orientation of the circuit coincide, otherwise, the sign is minus.*

For example, applying Kirchhoff's voltage law (KVL) to a network with the digraph of Figure 1.28 and the circuit $\{c, d, e\}$, (oriented as shown), the associated equation is $u_c(t) - u_d(t) + u_e(t) = 0$.

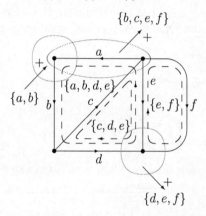

Figure 1.28. An oriented graph with associated incidence matrices.

Network Current Axiom: [Kirchhoff's current law] *The algebraic sum of current signals associated with the edges that belong to an oriented cutset, is equal to zero. By algebraic, we mean that every current appears in this sum with a plus sign if the orientation of the corresponding edge and the orientation of the cutset coincide. Otherwise, the sign is minus.*

For example, applying Kirchoff's current law (KCL) to the digraph of Figure 1.28 and the cutset $\{b, c, e, f\}$, (oriented as shown), the associated equation is $i_b(t) - i_c(t) + i_e(t) + i_f(t) = 0$.

The equations written by means of KVL and KCL, taken together, are also called *topological relations*. With every oriented cutset of an oriented network we have an associated linear equation called the *KCL equation* and with every oriented circuit of an oriented network we have an associated linear equation called the *KVL equation*.

The following proposition is intimately related to the Orthogonality Theorem.

Proposition 1.40 *Given a directed graph G with edge set E of cardinality b, let C be an oriented circuit and let S be an oriented cutset of G. Let p be the row matrix that describes the incidence of the circuit C against elements of E such that*

$$p_k = \begin{cases} 1 & \text{if } C \text{ contains the edge } k \text{ and the orientations of } C \text{ and } k \\ & \qquad coincide, \\ -1 & \text{if } C \text{ contains the edge } k \text{ and their orientations are opposite}, \\ 0 & \text{if } C \text{ does not contain the edge } k. \end{cases}$$

Let q be the row matrix that describes the incidence of the cutset S against elements of E in G such that

$$q_k = \begin{cases} 1 & \text{if } S \text{ contains the edge } k \text{ and the orientations of } S \text{ and } k \\ & \qquad coincide, \\ -1 & \text{if } S \text{ contains the edge } k \text{ and their orientations are opposite}, \\ 0 & \text{if } S \text{ does not contain the edge } k. \end{cases}$$

Suppose that columns in both matrices are taken in the same order. Then for every orientation of edges of G, and any orientation of C and S, the following scalar product is zero:

$$p^{\mathrm{T}} q = \sum_{k=1}^{b} p_k q_k = 0.$$

Proof According to the Orthogonality Theorem the intersection $C \cap S$ is of even cardinality. If $C \cap S = \emptyset$ then for each $k \in \{1, \ldots, b\}$, at least one of p_k and q_k is equal to zero. Therefore $\sum_{k=1}^{b} p_k q_k = 0$. Suppose now that $C \cap S$ is not empty. Then for an even nonzero number of edges k, both p_k and q_k are nonzero. In precisely half of these edges the orientations of C and S coincide in which case product $p_k q_k$ is equal to 1. For the other edges this product is -1 and hence again $\sum_{k=1}^{b} p_k q_k = 0$. □

For example, for the circuit $\{c, d, e\}$ and the cutset $\{b, c, e, f\}$ of the oriented graph of Figure 1.28 the matrices p and q are:

$$p = \begin{array}{c} \\ \{c, d, e\} \end{array} \begin{array}{cccccc} [a] & [b] & [c] & [d] & [e] & [f] \\ 0 & 0 & 1 & -1 & 1 & 0 \end{array},$$

$$q = \begin{array}{c} \\ \{b, c, e, f\} \end{array} \begin{array}{cccccc} [a] & [b] & [c] & [d] & [e] & [f] \\ 0 & 1 & -1 & 0 & 1 & 1 \end{array}.$$

So that $p^{\mathrm{T}} q = 0$.

Given a directed graph G with edge set E, let $\mathbf{C}'_o \subseteq \mathbf{C}_o$ be any collection of oriented circuits of G and let $\mathbf{S}'_o \subseteq \mathbf{S}_o$ be any collection of oriented cutsets of G, respectively. Let P' be the matrix that describes the incidence of members of $\mathbf{C}'_o$ against members of E of G. The entries for $i = 1, \ldots, |\mathbf{C}'_o|$ and $k = 1, \ldots, |E|$ are

$$p'_{ik} = \begin{cases} 1 & \text{if circuit } i \text{ contains edge } k \text{ and their orientations coincide,} \\ -1 & \text{if circuit } i \text{ contains edge } k \text{ and their orientations are opposite,} \\ 0 & \text{if circuit } i \text{ does not contain the edge } k. \end{cases}$$

Let Q' be the matrix that describes the incidence of members of $\mathbf{S}'_o$ against members of E of G. The entries for $j = 1, \ldots, |\mathbf{S}'_o|$ and $k = 1, \ldots, |E|$ are:

$$q'_{jk} = \begin{cases} 1 & \text{if cutset } i \text{ contains edge } k \text{ and their orientations coincide,} \\ -1 & \text{if cutset } i \text{ contains edge } k \text{ and their orientations are opposite,} \\ 0 & \text{if cutset } i \text{ does not contain the edge } k. \end{cases}$$

Thus both matrices P' and Q' are matrices with 1, -1 or 0 entries and we call them the *circuit incidence matrix* and the *cutset incidence matrix*, respectively. In matrix P' the rows correspond to oriented circuits and the columns to oriented edges, while in the matrix Q' the rows correspond to oriented cutsets and the columns to oriented edges.

Let P' and Q' be two matrices with the same number of columns whose entries are arbitrary real numbers. Denote by m' the number of rows of matrix P' and denote by n' the number of rows of matrix P'. It is said that P' and Q' are *orthogonal* if they satisfy the following relations:

$$P'(Q')^{\mathrm{T}} = 0_{m',n'} \text{ and } Q'(P')^{\mathrm{T}} = 0_{n',m'} \tag{o}$$

where $0_{m',n'}$ and $0_{n',m'}$ are $(m' \times n')$ and $(n' \times m')$ matrices respectively with zero entries.

The next corollary is an immediate consequence of Proposition 1.40:

Corollary 1.41 [to Proposition 1.40] *Let G be a directed graph with edge set E. Let P' be a matrix describing the incidence of a collection of oriented circuits $\mathbf{C}'_o \subseteq \mathbf{C}_o$ against the oriented edges of G and let Q' be a matrix describing the incidence of a collection of oriented cutsets $\mathbf{S}'_o \subseteq \mathbf{S}_o$ against the oriented edges of G. Suppose that columns in both matrices are taken in the same order. Then, P' and Q' are orthogonal.*

Let $\mathbf{C}'_o \subseteq \mathbf{C}_o$ be a collection of oriented circuits and let $\mathbf{S}'_o \subseteq \mathbf{S}_o$ be a collection of oriented cutsets of a graph G. Let P' and Q' be the associated incidence matrices whose columns are taken in the same order. Denote by u the $(b \times 1)$ column matrix whose components are voltages and denote by i the $(b \times 1)$ column matrix whose components are currents. Assume that rows in both are taken in the same order and that this order coincides with the order of columns of matrices P' and Q'. Then the set of equations based on Kirchhoff's laws written for a collection of circuits $\mathbf{C}'_o$ and a collection of cutsets $\mathbf{S}'_o$ can be described by the following matrix equations:

$$P'u = 0 \tag{a}$$

$$Q'i = 0. \tag{b}$$

For example, for the oriented graph of Figure 1.28 the circuit incidence matrix P' that corresponds to the collection of circuits $\mathbf{C}'_o = \{\{a, b, d, e\}, \{c, d, e\}, \{e, f\}\} \subseteq \mathbf{C}_o$ and the cutset incidence matrix Q' that corresponds to the collection of cutsets $\mathbf{S}'_o = \{\{a, b\}, \{b, c, e, f\}, \{d, e, f\}\} \subseteq \mathbf{S}_o$, are:

$$
P' =
\begin{array}{c}
\\
\{a,b,d,e\} \\
\{c,d,e\} \\
\{e,f\}
\end{array}
\begin{array}{cccccc}
[a] & [b] & [c] & [d] & [e] & [f] \\
1 & 1 & 0 & 1 & -1 & 0 \\
0 & 0 & 1 & -1 & 1 & 0 \\
0 & 0 & 0 & 0 & -1 & 1
\end{array},
$$

$$
Q' =
\begin{array}{c}
\\
\{a,b\} \\
\{b,c,e,f\} \\
\{d,e,f\}
\end{array}
\begin{array}{cccccc}
[a] & [b] & [c] & [d] & [e] & [f] \\
1 & -1 & 0 & 0 & 0 & 0 \\
0 & 1 & -1 & 0 & 1 & 1 \\
0 & 0 & 0 & -1 & -1 & -1
\end{array}.
$$

Obviously, P' and Q' are orthogonal. The total set of KCL equations is linearly dependent. So is the total set of KVL equations. In network topological analysis maximal collections of linearly independent KCL equations and maximal collections of linearly independent KVL equations are required. Many aspects of finding such collections will be considered in Chapter 2. As will be shown there, the cardinality of any maximal collection of linearly independent KCL equations is equal to the number of vertices of the graph associated with the network minus the number of 1-connected components of this graph. Also, the cardinality of any maximal collection of linearly independent KVL equations is equal to the number of edges of this graph minus the maximal number of linearly independent KCL equations. For example, for the digraph of Figure 1.28, the family of cutsets $\{\{a, b\}, \{b, c, e, f\}, \{d, e, f\}\}$ induces the set of KCL equations which is a maximal set of linearly independent KCL equations. For the same digraph, the family of circuits $\{\{a, b, d, e\}, \{c, d, e\}, \{e, f\}\}$ induces the set of KVL equations which is a maximal set of linearly independent KVL equations.

Because every network is closed by means of couplings, we can always treat the collection of its multiports as one *global multiport*. Suppose a network consists of b two-terminal ports, some of which are coupled forming a number of multiports. Then, the global multiport of a network has b ports and hence can be described by a constitutive relation $\mathcal{E}_b \subseteq S^{2b}$ of all $2b$-tuples of signals $(u_1(t), i_1(t); \ldots ; u_b(t), i_b(t))$ that are allowed on its ports. This set of $2b$-tuples includes all information concerning individual ports as well as their couplings. On the other hand, it contains no information concerning galvanic connections of network ports. The galvanic connections of a network are described by a special linear subspace of S^{2b}, denoted by $\mathcal{K}_b$, which we call *Kirchhoff's space*. The set $\mathcal{K}_b$ contains all $2b$-tuples $(u_1, i_1; \ldots ; u_b, i_b)$ that are simultaneously allowed by Kirchhoff's laws. In fact, $\mathcal{K}_b = \mathcal{U}_b \times \mathcal{I}_b$, where the set $\mathcal{U}_b$ consists of all b-tuples of voltage values $(u_1, \ldots , u_b)$ allowed by Kirchhoff's voltage laws and $\mathcal{I}_b$ consists of all b-tuples of current values $(i_1, \ldots , i_b)$, allowed by Kirchhoff's current laws. Clearly, both, $\mathcal{U}_b$ and $\mathcal{I}_b$ are linear subspaces of R^b over the field of real numbers R.

At first sight, it seems that the linear spaces $\mathcal{U}_b$ and $\mathcal{I}_b$ are mutually independent. But actually they are dependent in a sophisticated way. This dependence is captured by the following property:

Assertion 1.42 $\mathcal{U}_b$ *and* $\mathcal{I}_b$ *are mutually orthogonal in the sense that for any b-tuple* $(u_1, \ldots , u_b)$ *from* $\mathcal{U}_b$ *and any b-tuple* $(i_1, \ldots , i_b)$ *from* $\mathcal{I}_b$ *the sum* $\sum_{k=1}^{b} u_k i_k$ *is equal to zero.*

This property known as the *Theorem of Tellegen* (which will be proved in Chapter 2, Proposition 2.41), is directly connected with Proposition 1.40.

1.10   Bibliographic notes

The notions of circ and cut were first used by Reed and Maxwell [62, 41](who used the name seg instead of cut). The proofs of Assertions 1.30 and 1.31 can be found in Williams and Maxwell [76] and Chen [9]. For the proof of Assertion 1.35 see Tutte [69]. The proof of Assertion 1.37 can be found in Whitney [74]. For the proof of Assertion 1.39 see [70]. A general representation of affine multiports can be found for example in [48]. Kirchhoff's voltage and current laws were formally introduced in [35] by the German scientist G. Kirchhoff in the mid-nineteenth century (1847). Harary [26], Chen [8] Mayeda [42], Deo [14], Swammy and Thulasiraman [65], Gibbons [21] and Recski [61] provide further reading concerning graph theory while Desoer and Kuh [16], Fosseprez [17], Novak [50], Mathis [40], Newcomb [47], Hasler and Neirynck [30] and Recski [61] provide further reading concerning network theory.

2

Independence Structures

Independence is a unifying concept for linear algebra and graphs. A deep generalization of both through this unification, is contained in the notion of graphoids. Every graph can be seen as an interpretation of a graphoid in a particular 'coordinate system', called a 2-complete basis. From this prospect, a graphoid is an essential, coordinate free, geometrical notion for which each associated graph, if it exists, is just a particular view of the same generality. The concept of a graphoid can also be seen as a pair of set systems (dual matroids) whose members are called circuits and cutsets. The set of all circuits (cutsets) together with all their distinct unions we call a circ (cut) space. In this chapter, in the context of circuits and cutsets, we concentrate on two concepts of independence within graphs and graphoids. We first introduce independent collections of circuits and cutsets and then we use this concept to define independent edge subsets, that is, circuit-less and cutset-less subsets. From this point on, we take circuits and cutsets as primary notions. This material may be seen as a bridge between traditional graph theory and matroid theory. We also give a brief overview of properties of graphoids and methods in topological analysis of networks.

2.1 The graphoidal point of view

The space of all graphs can be divided into disjoint classes such that two graphs belong to the same class if they are 2-isomorphic. According to Assertion 1.37, chapter 1, with each such class we can uniquely associate a triple $(E, \mathbf{C}, \mathbf{S})$ where E is the common edge set of all graphs in the class and $\mathbf{C}$ and $\mathbf{S}$ are the common collections of circs and cuts respectively. For example, the pair of 2-isomorphic graphs of Figure 2.1 is a pair of distinct graphs and yet they have the same triple $(E, \mathbf{C}, \mathbf{S})$.

As we mentioned in chapter 1, the ring sum operation $\oplus$ is commutative and associative, the empty set is an identity element for the ring sum operation and every set is its own inverse. According to Propositions 1.11, 1.16 and

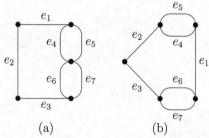

(a)   (b)

Figure 2.1. Two different graphs with the same cut and circ spaces.

1.20 of chapter 1, $\mathbf{C}$ and $\mathbf{S}$ are closed under the ring sum operation and any circuit and cutset always have an even number of edges in common. Thus, the following statements hold:

    (i) $(\mathbf{C}, \otimes \mid E)$ and $(\mathbf{S}, \otimes \mid E)$ are commutative groups,

    (ii) if $C \in \mathbf{C}$ and $S \in \mathbf{S}$, then $|C \cap S|$ is even.

    Conversely, let E be a finite nonempty set and let $\mathbf{C}$ and $\mathbf{S}$ be two collections of subsets of E satisfying (i) and (ii) (we call them the circ space and the cut space respectively). The members of $\mathbf{C}$ (respectively, $\mathbf{S}$) are called *circs* (*cuts*) and the triple $\mathcal{G} = (E, \mathbf{C}, \mathbf{S})$ is called a *binary graphoid*. The concept of a binary graphoid is a deep generalization of the concept of a graph. If $\mathcal{G} = (E, \mathbf{C}, \mathbf{S})$ is a binary graphoid, then $\mathcal{M} = (E, \mathbf{C})$ and $\mathcal{M}^* = (E, \mathbf{S})$ constitute the pair of *dual binary matroids* called the *circuit binary matroid* and the *cutset binary matroid*, respectively. The set E is called the *ground set* of the binary graphoid. As is discussed in the last two sections of this chapter, binary graphoids have a special hypergraph representation (hypergraphs whose edges are incident to an even nonzero number of vertices) and therefore it is justified to adopt the term *edge* to denote a member of the ground set of a binary graphoid.

    The minimal members of $\mathbf{C}$ (respectively, $\mathbf{S}$) are called *circuits* (*cutsets*). We denote the collection of minimal members of $\mathbf{C}$ (respectively, $\mathbf{S}$) by $\mathbf{C}_o$ ($\mathbf{S}_o$). Accordingly, no proper subset of a member of $\mathbf{C}_o$ ($\mathbf{S}_o$) is a member of $\mathbf{C}_o$ ($\mathbf{S}_o$). Given a set $\mathbf{C}_o$ ($\mathbf{S}_o$) of all circuits (cutsets), the corresponding set of circs (cuts) can be obtained by taking all unions of members of $\mathbf{C}_o$ ($\mathbf{S}_o$), including members of $\mathbf{C}_o$ ($\mathbf{S}_o$). Thus the triple $(E, \mathbf{C}_o, \mathbf{S}_o)$ generates $(E, \mathbf{C}, \mathbf{S})$ and hence $(E, \mathbf{C}_o, \mathbf{S}_o)$ is another representation of a binary graphoid.

    The following statements are equivalent:

    (0) $\mathcal{M}$ *is a circuit binary matroid.*

    (1) *For any two distinct circuits C' and C'' of $\mathcal{M}$, $C' \oplus C''$ is either a circuit or a union of disjoint circuits* (see also Corollary 1.17, chapter 1).

    (2) *For any two distinct circuits C' and C'' of $\mathcal{M}$, and any two members x and y of $C' \cap C''$, there is a circuit $C \subseteq (C' \cup C'') \setminus \{x, y\}$* (see also Remark 1.19 of chapter 1).

Binary graphoids could be embedded into the larger class of general graphoids. The motivation for introducing general graphoids can be found in several statements about graphs within the text (Corollaries 1.13 and 1.18 of chapter 1 and the Orthogonality Theorem). Consider a triple $(E, \mathbf{C}_o, \mathbf{S}_o)$ that satisfies the following conditions:

(S1) No proper subset of a member of $\mathbf{S}_o$ is a member of $\mathbf{S}_o$.

(S2) If S_1 and S_2 are distinct members of $\mathbf{S}_o$ and $x \in S_1 \cap S_2$ then for each $y \in S_1 \setminus S_2$ there exists a member of S of $\mathbf{S}_o$ such that $y \in S \subseteq (S_1 \cup S_2) \setminus \{x\}$,

(C1) No proper subset of a member of $\mathbf{C}_o$ is a member of $\mathbf{C}_o$.

(C2) If C_1 and C_2 are distinct members of $\mathbf{C}_o$ and $x \in C_1 \cap C_2$ then for each $y \in C_1 \setminus C_2$ there exists a member of C of $\mathbf{C}_o$ such that $y \in C \subseteq (C_1 \cup C_2) \setminus \{x\}$,

(O) If $C \in \mathbf{C}_o$ and $S \in \mathbf{S}_o$, then $|C \cap S| \neq 1$.

The same statements hold for binary graphoids. The converse generally is not true. That is, if $\mathbf{C}_o$ and $\mathbf{S}_o$ are collections of subsets of a finite set of E satisfying conditions (S1), (S2), (C1), (C2) and (O), the triple $(E, \mathbf{C}_o, \mathbf{S}_o)$ might not be a binary graphoid. A triple $(E, \mathbf{C}_o, \mathbf{S}_o)$ that satisfies (S1), (S2), (C1), (C2) and (O) we take as axioms of a *general graphoid*.

Conditions (C1) and (C2) define a notion called a *matroid* (*circuit matroid*). Similarly, conditions (S1) and (S2) also define a *matroid* (*cutset matroid*). Circuit and cutset matroids are dual matroids by means of condition (O) which is the Orthogonality Theorem. Instead of condition (O) duality can be also described by any of the following conditions:

(D) the complement of a maximal subset of E which does not contain any member of $\mathbf{C}_o$ is a maximal subset of E which does not contain any member of $\mathbf{S}_o$ and vice versa.

Thus, a graphoid can be seen as a pair of dual matroids $\mathcal{M} = (E, \mathbf{C}_o)$ and $\mathcal{M}^* = (E, \mathbf{S}_o)$. The following statements are equivalent:

(i) $\mathcal{M}^* = (E, \mathbf{S}_o)$ *is dual to* $\mathcal{M} = (E, \mathbf{C}_o)$.

(ii) *Assume that* $S \subseteq E$ *and* $S \neq \emptyset$. *Then* $S \in \mathbf{S}_o$ *iff* S *is a minimal nonempty subset of* E *such that* $|C \cap S| \neq 1$ *for all* $C \in \mathbf{C}_o$.

(iii) *Assume that* $B \subseteq E$. *Then* B *is a maximum subset of* E *which does not contain any member of* $\mathbf{C}_o$ *iff* $B^* = E \setminus B$ *is a maximum subset of* E *which does not contain any member of* $\mathbf{S}_o$.

The set of all circuits of a graph forms the so-called *graphic matroid* and the set of all cutsets of a graph forms the so-called *cographic matroid*. However, not every matroid $\mathcal{M} = (E, \mathbf{C}_o)$ is graphic.

Replacement of conditions (S2) and (C2) with the Painting Theorem results in an equivalent system of axioms of a general graphoid:

(S1) No proper subset of a member of $\mathbf{S}_o$ is a member of $\mathbf{S}_o$.

(C1) No proper subset of a member of $\mathbf{C}_o$ is a member of $\mathbf{C}_o$.

(P) For any partition $(E_1, \{e\}, E_2)$ of the set E there is either a circuit

$C \in \mathbf{C}_o$ containing e and no elements of E_2 or a cutset $S \in \mathbf{S}_o$ containing e and no elements of E_1.

(O) If $C \in \mathbf{C}_o$ and $S \in \mathbf{S}_o$, then $|C \cap S| \neq 1$.

Let $\mathcal{G} = (E, \mathbf{C}_o, \mathbf{S}_o)$ be a binary graphoid where $\mathbf{C}_o$ is its set of circuits and $\mathbf{S}_o$ is its set of cutsets and let A be a subset of E. Let $\mathbf{C}'_o(A) = \{C' \in \mathbf{C}_o \mid C' \subseteq A\}$ and $\mathbf{S}'_o(A) = \{S' \mid S' = S \cap A, S \in \mathbf{S}_o\}$. It is clear that $(A, \mathbf{C}'_o(A), \mathbf{S}'_o(A))$ is a binary graphoid defined on A. We denote this binary graphoid by $\mathcal{G} \mid A$ and call it the *reduction* of the binary graphoid $\mathcal{G}$ from E to A. Let $\mathbf{C}''_o(A) = \{C'' \mid C'' = C \cap A, C \in \mathbf{C}_o\}$ and $\mathbf{S}''_o(A) = \{S'' \in \mathbf{S}_o \mid S'' \subseteq A\}$. It is clear that $(A, \mathbf{C}''_o(A), \mathbf{S}''_o(A))$ is also a binary graphoid defined on A. We denote this binary graphoid by $\mathcal{G} \cdot A$ and call it the *contraction* of the binary graphoid $\mathcal{G}$ from E to A.

2.2 Independent collections of circs and cuts

In what follows we focus on graphs and binary graphoids. Let $\mathbf{F}$ be a collection of subsets of a finite nonempty set E closed with respect to the set operation ring sum $\oplus$. Since $\oplus$ is both an associative and a commutative operation and since for any member Q of $\mathbf{F}$ the following relations hold: $Q \oplus \emptyset = Q$, $\emptyset \oplus Q = Q$ and $Q \oplus Q = \emptyset$, it follows that the pair $(\mathbf{F}, \oplus)$ is a commutative group which we call a *group set system* on the ground set E. We say that a collection $\mathbf{Q} \subseteq \mathbf{F}$ is an independent collection of members of a group set system $\mathbf{F}$ if there is no subcollection $\{Q'_1, \ldots, Q'_p\}$, of $\mathbf{Q}$ such that $Q'_1 \oplus \cdots \oplus Q'_p = \emptyset$. Otherwise, we say that $\mathbf{Q}$ is a dependent collection of members of a group set system. Obviously, any subcollection of an independent collection is also independent. An independent collection $\mathbf{Q}$ of members of $\mathbf{F}$ is a *maximal independent collection* if no other independent collection of members of $\mathbf{F}$ contains $\mathbf{Q}$ as a proper subcollection. We define the ring sum $\oplus$ of two collections $\mathbf{X}$ and $\mathbf{Y}$ of members of $\mathbf{F}$ as $\mathbf{X} \oplus \mathbf{Y} = \{X \oplus Y \mid X \in \mathbf{X}, Y \in \mathbf{Y}\}$. Given a collection $\mathbf{Q} = \{Q_1, \ldots, Q_n\}$ of members of $\mathbf{F}$, we define $\langle \mathbf{Q} \rangle = \{\emptyset, Q_1\} \oplus \cdots \oplus \{\emptyset, Q_n\}$. It is easy to see that $\langle \mathbf{Q} \rangle$, as a collection of members of $\mathbf{F}$, is closed with respect to $\oplus$ and therefore $(\langle \mathbf{Q} \rangle, \oplus)$ is a subgroup of $(\mathbf{F}, \oplus)$. If $\langle \mathbf{Q} \rangle = \mathbf{F}$, then we say that $\mathbf{Q}$ is a *generator of a group set system* $\mathbf{F}$.

Lemma 2.1 *Let $\mathbf{F}$ be a collection of subsets of a finite nonempty set E, closed with respect to the ring sum operation, and let $\mathbf{Q}$ be an independent collection of nonempty members of $\mathbf{F}$. Suppose Q_0 and Q_1 are two distinct members of $\mathbf{Q}$. Then the collection $\mathbf{Q}'$, obtained from $\mathbf{Q}$ by replacing Q_1 by $Q_0 \oplus Q_1$, is also an independent collection of nonempty members of $\mathbf{F}$.*

Proof The ring sum of any number of distinct members of $\mathbf{Q}'$ is at the same time the ring sum of some distinct members of $\mathbf{Q}$. Consequently, the ring sum of a number of distinct members of $\mathbf{Q}'$ is nonempty iff the same holds for the ring sum of members of $\mathbf{Q}$. But $\mathbf{Q}$ is an independent collection of nonempty members of $\mathbf{F}$ and thus so is $\mathbf{Q}'$.   □

Lemma 2.2 *Let* $\mathbf{F}$ *be a collection of subsets of a finite nonempty set* E, *closed with respect to the ring sum. Then, any maximal independent collection of members of* $\mathbf{F}$ *generates* $\mathbf{F}$.

Proof Let $\mathbf{Q} = \{Q_1, \ldots, Q_m\}$ be a maximal independent collection of members of $\mathbf{F}$ and suppose there exists $Q_o \in \mathbf{F}, Q_o \neq \emptyset$ that does not belong to $\langle \mathbf{Q} \rangle$. Then for any subcollection $\mathbf{Q}' = \{Q'_1, \ldots, Q'_p\}$ of $\mathbf{Q}$, $Q'_1 \oplus Q'_2 \oplus \cdots \oplus Q'_p \neq Q_o$. Thus $(Q_1 \oplus Q_2 \oplus \cdots \oplus Q_m) \oplus Q_o \neq \emptyset$. This, implies that $\mathbf{Q}'' = \mathbf{Q} \cup \{Q_o\}$ is an independent collection of $\mathbf{F}$ which contradicts the assumption that $\mathbf{Q}$ is a maximal independent collection of $\mathbf{F}$. Hence $Q_o \in \mathbf{Q}$ and consequently any maximal independent collection of $\mathbf{F}$ generates $\mathbf{F}$.   □

Proposition 2.3 *All maximal independent collections of members of a group set system have the same cardinality.*

Proof The proof is by contradiction. Let $\mathbf{Q} = \{Q_1, \ldots, Q_n\}$ and $\mathbf{P} = \{P_1, \ldots, P_m\}$ be two maximal independent collections of members of $\mathbf{F}$ and suppose that $|\mathbf{P}| = m < n = |\mathbf{Q}|$. According to Lemma 2.2, $\langle \mathbf{Q} \rangle = \langle \mathbf{P} \rangle = \mathbf{F}$, that is, each of $\mathbf{P}$ and $\mathbf{Q}$ generates $\mathbf{F}$. On the other hand, because $\langle \mathbf{P} \rangle = \{\emptyset, P_1\} \oplus \cdots \oplus \{\emptyset, P_m\} = F(G) \supset \mathbf{Q}$ we deduce that $Q_k = F_{k_1} \oplus \cdots \oplus F_{km}$, where $F_{ks} \in \{\emptyset, P_s\}$ for $k \in \{1, \ldots, n\}$. Thus we have a system of n relations that express each Q_k in terms of P_s, $s = 1, \ldots, m$. Now consider a sequence of consecutive equivalent transformations of the system of relations $\{Q_k = F_{k_1} \oplus \cdots \oplus F_{km} \mid k = 1, \ldots, n\}$, in a manner analogous to the well-known reduction to row echelon form in linear algebra. By an equivalent transformation we mean a transformation which preserves independence in the sense that any new collection obtained $\mathbf{Q}' = \{Q'_1, \ldots, Q'_n\}$ is still an independent collection. The procedure is carried out in a number of steps as follows.

Step 0. Set $M \leftarrow \{1, \ldots, m\}$, $i \leftarrow 0$, and for $k \in \{1, \ldots, n\}$, $Q_k^0 \leftarrow Q_k$.

Step 1. Choose some $j \in M$ and find the subset K_j of $\{1, \ldots, n\}$ such that for all $k \in K_j$, $F_{kj} = P_j$. Notice that K_j is always nonempty. Let $p \in K_j$ be a particular member of K_j. If $K_j \setminus \{p\}$ is empty, go to *Step 2*. Otherwise, that is if $K_j \setminus \{p\}$ is nonempty, consider the new system of relations $\{Q_k^{i+1} = F_{k1}^{i+1} \oplus \cdots \oplus F_{km}^{i+1} \mid k \in \{1, \ldots, n\}\}$ where $Q_k^{i+1} = Q_k^i$, for $k \in \{1, \ldots, n\} \setminus (K_j \setminus \{p\})$ and $Q_k^{i+1} = Q_k^i \oplus Q_p^i$, for $k \in K_j \setminus \{p\}$. By construction $F_{kj}^{i+1} = P_j$ for $k = p$ and $F_{kj}^{i+1} = \emptyset$ for $k \neq p$. According to Lemma 2.1 the collection $\{Q_1^{i+1}, \ldots, Q_n^{i+1}\}$ is independent if $\{Q_1^i, \ldots, Q_n^i\}$ is independent. Set $i \leftarrow i + 1$ and go to *Step 2*.

Step 2. Set $M \leftarrow M \setminus \{j\}$. If M is nonempty, go to *Step 1*. Otherwise go to *Step 3*.

Step 3. Stop.

By repeating the process $s(\leq m)$ times we obtain the system of relations $\{Q_k^s = F_{k_1}^s \oplus \cdots \oplus F_{km}^s \mid k \in \{1, \ldots, n\}\}$ where no two relations have the same members of $\mathbf{P}$ and no two members of $\mathbf{P}$ appear in the same relation. Because $|\mathbf{P}| = m < n = |\mathbf{Q}|$, there is at least one $k \in \{1, \ldots, n\}$ for which $Q_k^s = \emptyset$, that is, $Q_a^{s-1} \oplus \cdots \oplus Q_p^{s-1} = \emptyset$. This means that the collection $\mathbf{Q}^{s-1} = \{Q_1^{s-1}, \ldots, Q_n^{s-1}\}$ is not independent which implies that $\mathbf{Q}$ also is not independent; a contradiction.   □

To illustrate the procedure of consecutive transformations obtained in the above proof, consider the following example of a system of relations in which $m = 3$ and $n = 4$:

$$Q_1 = P_1 \oplus \emptyset \oplus P_3,$$
$$Q_2 = P_1 \oplus P_2 \oplus \emptyset,$$
$$Q_3 = \emptyset \oplus \emptyset \oplus P_3,$$
$$Q_4 = P_1 \oplus P_2 \oplus P_3.$$

To eliminate P_1 we combine the first relation with the second and the fourth:

$$Q_1' = Q_1 = P_1 \oplus \emptyset \oplus P_3,$$
$$Q_2' = Q_1 \oplus Q_2 = \emptyset \oplus P_2 \oplus P_3,$$
$$Q_3' = Q_3 = \emptyset \oplus \emptyset \oplus P_3,$$
$$Q_4' = Q_1 \oplus Q_4 = \emptyset \oplus P_2 \oplus \emptyset.$$

To eliminate P_2 we combine the second relation with the fourth:

$$Q_1'' = Q_1' = Q_1 = P_1 \oplus \emptyset \oplus P_3,$$
$$Q_2'' = Q_2' = Q_1 \oplus Q_2 = \emptyset \oplus P_2 \oplus P_3,$$
$$Q_3'' = Q_3' = Q_3 = \emptyset \oplus \emptyset \oplus P_3,$$
$$Q_4'' = Q_2' \oplus Q_4' = \emptyset \oplus \emptyset \oplus P_3.$$

To eliminate P_3 we combine the third relation with the fourth:

$$Q_1''' = Q_1'' = Q_1' = Q_1 = P_1 \oplus \emptyset \oplus P_3,$$
$$Q_2''' = Q_2'' = Q_2' = Q_1 \oplus Q_2 = \emptyset \oplus P_2 \oplus P_3,$$
$$Q_3''' = Q_3'' = Q_3' = Q_3 = \emptyset \oplus \emptyset \oplus P_3,$$
$$Q_4''' = Q_3'' \oplus Q_4'' = Q_2' \oplus Q_3' \oplus Q_4' = (Q_1 \oplus Q_2) \oplus Q_3 \oplus (Q_1 \oplus Q_4) = \emptyset \oplus \emptyset \oplus \emptyset \oplus \emptyset = \emptyset.$$

Finally, we observe that because $(Q_1 \oplus Q_2) \oplus Q_3 \oplus (Q_1 \oplus Q_4) = Q_2 \oplus Q_3 \oplus Q_4$, the last equation becomes $Q_2 \oplus Q_3 \oplus Q_4 = \emptyset$, which contradicts the assumption that $\mathbf{Q}$ is an independent collection of circs.

If we interpret $\mathbf{F}$ as the circ space $\mathbf{C}$ (respectively, cut space $\mathbf{S}$) of a binary graphoid $\mathcal{G}$ then, from Lemma 2.2, and Proposition 2.3, it follows that any maximal independent subcollection of $\mathbf{C}$ ($\mathbf{S}$) generates $\mathbf{C}$ ($\mathbf{S}$). Moreover the following holds: *All maximal independent collections of circs (cuts) of binary matroid have the same cardinality.*

Notice also that the statement of Lemma 2.1 still holds if instead of independent collections we consider maximal independent collections.

Corollary 2.4 [to Proposition 2.3] *Given a binary graphoid $\mathcal{G} = (E, \mathbf{C}_o, \mathbf{S}_o)$ let A be a subset of E. Then all maximal independent collections of circs (cuts) of $\mathcal{G}$, whose edges belong entirely to A, have the same cardinality.*

Proof Let $G \mid A$ $(G \cdot A)$) be the restriction (contraction) of $\mathcal{G}$ from E to A. It is clear that a circ (cut) of $\mathcal{G}$ entirely belongs to A iff it is a circ (cut) of $\mathcal{G} \mid A$ $(\mathcal{G} \cdot A)$). Moreover a collection of circs (cuts) of $\mathcal{G}$ is a maximal independent collection of circs (cuts) of $\mathcal{G}$ whose members belong entirely to A iff it is a maximal independent collection of circs (cuts) of $\mathcal{G} \mid A$ $(\mathcal{G} \cdot A)$. But, according to Proposition 2.3, all maximal independent collections of circs (cuts) of $\mathcal{G} \mid A$ $(\mathcal{G} \cdot A)$) have the same cardinality, and so the result follows. $\square$

Given a collection $\mathbf{A} \subseteq \mathbf{C}$ $(\mathbf{B} \subseteq \mathbf{S})$ of circs (cuts) of a binary graphoid $\mathcal{G} = (E, \mathbf{C}, \mathbf{S})$ suppose that there is a bijection f from $\mathbf{A}$ $(\mathbf{B})$ to a subset A (B) of the set E such that for every $C \in \mathbf{A}$ $(S \in \mathbf{B})$, $f(C)$ $(f(S))$ belongs to $C \cap A$ $(S \cap B)$ and does not belong to any other member of the collection $\mathbf{A}$ $(\mathbf{B})$. Then, we say that a collection of circs $\mathbf{A} \subseteq \mathbf{C}$ and a subset $A \subseteq E$ (respectively, a collection of cuts $\mathbf{B} \subseteq \mathbf{S}$ and a subset $B \subseteq E$) are *in correspondence*. Clearly, because we have a bijection, $|A| = |\mathbf{A}|$ $(|B| = |\mathbf{B}|)$.

Proposition 2.5 *If a collection $\mathbf{A} \subseteq \mathbf{C}$ $(\mathbf{B} \subseteq \mathbf{S})$ of circs (cuts) of a binary graphoid $\mathcal{G} = (E, \mathbf{C}, \mathbf{S})$ is in correspondence with a subset A (B) of E, then A (B) is an independent collection of circs (cuts).*

Proof The proof is by contradiction. Suppose that the collection $\mathbf{A} \subseteq \mathbf{C}$ $(\mathbf{B} \subseteq \mathbf{S})$ of circs (cuts) of a binary graphoid $\mathcal{G}$ which is in correspondence with the subset A (B) of E is not independent. Then, there is a subcollection $\{C_1, \ldots, C_p\}$ $(\{S_1, \ldots, S_q\})$ of circs (cuts) of $\mathbf{A}$ $(\mathbf{B})$ such that $C_1 \oplus \cdots \oplus C_p = \emptyset$ $(S_1 \oplus \cdots \oplus S_q = \emptyset)$. Since each circ (cut) of the collection contains an edge of E that does not belong to any other member of the collection, we conclude that $C_1 \oplus \cdots \oplus C_p$ $(S_1 \oplus \cdots \oplus S_q)$ must contain a nonempty subset of A (B), a contradiction. $\square$

The converse of Proposition 2.5 is generally not true. As a counter example consider the independent collection $\{\{e_3, e_4, e_5\}, \{e_5, e_6, e_7\}, \{e_2, e_4, e_6, e_7\}\}$ of circs of the graph of Figure 1.1 (Chapter 1) and the independent collection $\{\{e_2, e_3, e_5, e_6\}, \{e_2, e_3, e_5, e_7\}, \{e_4, e_5, e_6\}\}$ of cuts of the same graph.

We use the phrase 'circless (cutless) set of edges' to mean a set of edges which does not contain a circ (cut) as a subset. Similarly, the phrase 'circuit-less (cutset-less) set of edges' means a set of edges which does not contain a circuit (cutset) as a subset. A subset of edges is circless (cutless) iff it is circuit-less (cutset-less).

The following proposition provides a necessary and sufficient condition for a subset of the ground set of a binary graphoid to be in correspondence with an independent collection of its circs (cuts).

Proposition 2.6 *A subset A (B) of the ground set E of a binary graphoid is in correspondence with an independent collection of its circs (cuts) of the binary graphoid iff A (B) is a cutless (circless) subset of the ground set of the binary graphoid.*

Proof Let A (B) be a subset of the ground set E of a binary graphoid which is in correspondence with an independent collection **A** (**B**) of circs (cuts) and suppose that A (B) contains a cut S (a circ C) as a subset. Denote by $\mathbf{A}_S$ ($\mathbf{B}_C$) the subcollection of **A** (**B**) in correspondence with the cut $S \subseteq A$ (circ $C \subseteq B$). Since each edge in S (C) belongs to exactly one member of $\mathbf{A}_S$ ($\mathbf{B}_C$), we conclude that each member of $\mathbf{A}_S$ ($\mathbf{B}_C$) has exactly one edge in common with S (C). But, from the Orthogonality Theorem, the intersection of a circ and a cut has even cardinality, and so we have a contradiction.

Conversely, let A (B) be a cutless (circless) subset of the ground set E of a binary graphoid. Then, according to the Painting Theorem, each member of A (B) forms a circuit (cutset) with elements in $E \setminus A$ ($E \setminus B$) only. Denote by $C(x)$, $(S(y))$ the circuit (cutset) that an edge $x \in A$ ($y \in B$) forms with elements in $E \setminus A$ ($E \setminus B$) only and denote by **A** (**B**) the collection $\{C(x) \mid x \in A\}$ ($\{S(x) \mid y \in B\}$). Clearly **A** (**B**) is in correspondence with A (B) and hence, according to Proposition 2.5, **A** (**B**) is an independent collection of circs (cuts). □

The next proposition connects maximal independent collections of circs (cuts) with maximal cutset-less (circuit-less) subsets of the ground set of a binary graphoid.

Proposition 2.7 *A subset A (B) of the ground set E of a binary graphoid is in correspondence with a maximal independent collection of circs (cuts) of the binary graphoid iff A (B) is a maximal cutless (circless) subset of E.*

Proof Let A (B) be a subset of the ground set E in correspondence with a maximal independent collection **A** (**B**) of circs (cuts). From Proposition 2.6, A (B) is a cutless (circless) set. Suppose that A (B) is not a maximal cutless (circless) subset, that is, suppose that there is a superset A' (B') of A (B) which is cutless (circless). According to Proposition 2.6, there is an independent collection of circs (cuts) $\mathbf{A}'$ ($\mathbf{B}'$) in correspondence with A' (B'). Since A (B) is a proper subset of A' (B'), it follows that $|A| < |A'|$ ($|B| < |B'|$). But A (B) has the same cardinality as **A** (**B**) and A' (B') has the same cardinality as $\mathbf{A}'$ ($\mathbf{B}'$) and hence, the cardinality of **A** (**B**) is less than the cardinality of $\mathbf{A}'$ ($\mathbf{B}'$). This contradicts the assumption that **A** (**B**) is a maximal independent collection of circs (cuts). Consequently, A (B) is a maximal cutless (circless) subset.

Conversely, let A (B) be a maximal cutless (circless) subset of the ground set E of a binary graphoid. Since A (B) is cutless (circless), according to

Proposition 2.6 it is in correspondence with an independent collection of circs (cuts) **A** (**B**). Suppose that **A** (**B**) is not a maximal collection of independent circs (cuts), that is, suppose that there is a superset **A'** (**B'**) of **A** (**B**) which is also an independent collection of circs (cuts). Then, according to Proposition 2.6, there is a circless (cutless) subset A' (B') of E in correspondence with **A'** (**B'**). Since **A** (**B**) is a proper subset of **A'** (**B'**), it follows that $|A| < |A'|$ ($|B| < |B'|$) which contradicts the assumption that A (B) is a maximal cutless (circless) subset of E. Consequently, **A** (**B**) is a maximal collection of independent circs (cuts).   □

Using dual arguments to those in the proof of Proposition 2.7 we can prove that the following statement also holds:

Corollary 2.8 [to Propositions 2.7] *A collection of all circs (cuts) is in correspondence with a maximal cutless (circless) subset of the ground set of a binary graphoid iff it is a maximal independent collection of circs (cuts).*

Proposition 2.9 *Among all maximal independent collections of circuits (cutsets) of a binary graphoid G at least one is in correspondence with a subset of the ground set of G.*

Proof Suppose there is at least one nonempty independent collection of circuits (cutsets) of a binary graphoid G. Then at least one edge in E is not a self-cutset edge and therefore there is a nonempty cutless (circless) subset of E. Let A (B) be a maximal cutless (circless) subset of the ground set E of a binary graphoid G. Then, according to the Painting Theorem, each member of A (B) forms a circuit (cutset) with edges in $E \setminus A$ ($E \setminus B$) only. Denote by $C(x)$, ($S(y)$) the circuit (cutset) that an edge $x \in A$ ($y \in B$) forms with edges in $E \setminus A$ ($E \setminus B$) only and denote by **A** (**B**) the collection $\{C(x) \mid x \in A\}$ ($\{S(x) \mid y \in B\}$). Clearly **A** (**B**) is in correspondence with A (Y) and hence, according to Proposition 2.5, **A** (**B**) is an independent collection of circs (cuts). Since A (B) is a maximal cutless (circless) subset, it follows from Proposition 2.6 that **A** (**B**) is a maximal independent collection of circs (cuts). Notice again that **A** (**B**) is in correspondence with A (B).   □

The next two statements are immediate consequences of Proposition 2.7 and Proposition 2.9.

Corollary 2.10 [to Propositions 2.7 and 2.9] *The cardinality of a maximal independent collection of circs (cuts) of a binary graphoid G is less or equal to the cardinality of the ground set of G.*

Corollary 2.11 [to Proposition 2.7 and 2.9] *Let A be a subset of the ground set of a binary graphoid G. Then A (B) is a maximal cutset-less (circuit-less) subset of A iff there is a maximal independent collection of circuits (cutsets) of G that is in correspondence with A (B).*

The following procedure converts a maximal collection of nonempty circs (cuts) that is not in correspondence with a subset of the ground set of a binary graphoid into a maximal independent collection of circs (cuts) which is in correspondence. Given a binary graphoid $\mathcal{G}$ let $\mathbf{C} = \{C_1, \ldots, C_p\}$ (respectively, $\mathbf{S} = \{S_1, \ldots, S_q\}$) be a maximal collection of nonempty circuits (cutsets) that is not in correspondence with a subset of ground set of $\mathcal{G}$.

Step 0. Set $i \leftarrow 0$, $(j \leftarrow 0,)$ $C_k^0 \leftarrow C_k$, for $k \leq p$ $(S_l^0 \leftarrow S_l$, for $l \leq q)$.

Step 1. Take C_{i+1}^i (S_{j+1}^j) of $\mathbf{C}^i$ $(\mathbf{S}^j)$ and mark an edge x_{i+1} (y_{j+1}) in it. Because no circuit (cutset) of a binary graphoid is empty, x_{i+1} (y_{j+1}) obviously exists. Replace each circuit C (cutset S) in $\mathbf{C}^i \setminus \{C_{i+1}^i\}$ $(\mathbf{S}^j \setminus \{S_{j+1}^j\})$ that contains x_{i+1} (y_{j+1}) by $C_{i+1}^i \oplus C$ $(S_{j+1}^j \oplus S)$. From Lemma 2.1, the collection obtained $\mathbf{C}^{i+1} = \{C_1^{i+1}, \ldots, C_p^{i+1}\}$ $(\mathbf{S}^{j+1} = \{S_1^{j+1}, \ldots, S_q^{j+1}\})$ is still independent. But $\mathbf{C}^{i+1}$ $(\mathbf{S}^{j+1})$ has the same cardinality p (q) as $\mathbf{C}$ $(\mathbf{S})$ does and hence, according to Proposition 2.3, $\mathbf{C}^{i+1}$ $(\mathbf{S}^{j+1})$ is also maximal. Notice that for $1 \leq k \leq i$, C_k^{i+1} (for $1 \leq l \leq j$, S_l^{j+1}) is the only member of $\mathbf{C}^{i+1}$ $(\mathbf{S}^{j+1})$ that contains the edge x_k (y_l).

Step 2. If $i \leq p - 1$ $(j \leq q - 1)$ set $i \leftarrow i + 1$ $(j \leftarrow j + 1)$ and go to *Step 1*. Otherwise go to *Step 3*.

Step 3. Stop.

Notice that using the above procedure we obtain a collection $\mathbf{C}^m$ $(\mathbf{S}^n)$ in which each of p (q) members contains an edge that does not belong to any other member of the collection. Clearly, $m = \min\{p, |E|\}$ (respectively, $n = \min\{q, |E|\}$). Since $\mathbf{C}^m$ $(\mathbf{S}^n)$ is a maximal independent collection of circuits (cutsets), according to Corollary 2.10, it follows that $|E| \geq p$ $(|E| \geq q)$. Hence $m = p$ $(n = q)$.

2.3   Maximal circless and cutless sets

In the next two sections our material will be expressed in graph-theoretic terms. However the material is general in the sense that it can be easily extended from graphs to binary graphoids (the only exception is Remark 2.21).

Let E' be a subset of edges of a graph G. For a circless (cutless) set A (B) that entirely belongs to E', we say that it is a *maximal circless (maximal cutless)* subset of E' if no circless A' exists (no cutless B' exists) such that $A \subset A' \subseteq E'$ $(B \subset B' \subseteq E')$. The cardinality of a maximal circless (cutless) subset of an edge subset E' of a graph is called the *rank (corank)* of E'.

Proposition 2.12 *Let E' be a subset of edges of a graph G. All maximal circless (cutless) subsets of E' have the same cardinality.*

Proof According to Corollary 2.11 for each maximal circless (cutless) subset of E' there is an independent collections of circs (cuts) in correspondence with it. Since all maximal independent collections of circs (cuts), whose members belong entirely to E', have the same cardinality (Corollary 2.4), it follows that all maximal circless (cutless) subsets of E' also have the same cardinality. □

The following corollary describes a collective property of circless subsets of a graph.

Corollary 2.13 [to Proposition 2.12] *Let* **I** *be the collection of all circuit-less (cutset-less) subsets of a graph, then the following conditions are satisfied.*

(I1) $\emptyset \in \mathbf{I}$.

(I2) *If* $X \in \mathbf{I}$ *and* $Y \subseteq X$ *then* $Y \in \mathbf{I}$.

(I3) *If* $X, Y \in \mathbf{I}$ *and* $|X| < |Y|$, *then there exists* $Z \in \mathbf{I}$ *such that* $X \subset Z \subseteq X \cup Y$.

Proof (I1) follows from the following facts: each circuit (cutset) is nonempty and the only subset of the empty set is the empty set itself.

(I2) trivially true.

(I3) Let $X, Y \in \mathbf{I}$ and $|X| < |Y|$ and suppose there is no $Z \in \mathbf{I}$ such that $X \subset Z \subseteq X \cup Y$. This means that X is a maximal circuit-less (cutset-less) subset of the union $X \cup Y$. But Y is a circuit-less (cutset-less) subset of the graph and consequently, it is also a circuit-less (cutset-less) subset of $X \cup Y$. Since according to Proposition 2.12 all maximal circuit-less (cutset-less) subsets of $X \cup Y$ have the same cardinality, it follows that $|X| \geq |Y|$ which contradicts our assumption that $|X| < |Y|$. □

Proposition 2.14 *Let* E' *be a subset of edges of a graph* G *and let* X (Y) *be a maximal circuit-less (cutset-less) subset of* E'. *Then each edge in* $E' \setminus X$ $(E' \setminus Y)$ *forms a unique circuit (cutset) with edges in* X (Y) *only.*

Proof Let E' be a subset of edges of a graph G and let X (Y) be a circuit-less (cutset-less) subset of E'. Let x (y) be an edge in $E' \setminus X$ $(E' \setminus Y)$, then $X \cup \{x\}$ $(Y \cup \{y\})$ contains at least one circuit (cutset) of G. Clearly, if there is more than one circuit then each circuit (cutset) must contain the edge x (y). Suppose two such circuits are C_1 and C_2 (cutsets S_1 and S_2), then $C_1 \oplus C_2$ $(S_1 \oplus S_2)$ is a circ (cut) that does not contain the edge x (y), that is, consists of edges in X (Y) only. This contradicts the assumption that X (Y) is circuit-less (cutset-less) and thus x (y) forms a unique circuit (cutset) with edges in X (Y) only. □

A circuit (cutset) that an edge in the complement of a circuit-less (cutset-less) subset X (Y) of edges forms with edges in X (Y) only, is called a *fundamental circuit (fundamental cutset)* with respect to X (Y).

The following corollary is well known.

Corollary 2.15 [to Propositions 2.7 and 2.12] *Let X (Y) be a maximal circuit-less (cutset-less) subset of edges of a graph G, then the collection of all fundamental circuits (cutsets) with respect to X (Y) is a maximal independent collections of circuits (cutsets).*

Given the set $\mathbf{C}$ of all circs of a graph, we can find a maximal circless subset of edges. From this maximal circless subset of edges we can find a set of fundamental cutsets and this may be used to generate $\mathbf{S}$, the set of all cuts of the graph. Thus $\mathbf{S}$ may in principle be found from $\mathbf{C}$. In a completely dual manner, $\mathbf{C}$ may be found from $\mathbf{S}$.

The next proposition describes an important property of maximal circless (and cutless) subsets of a graph.

Proposition 2.16 [the Fundamental Exchange Operation Theorem] *Let X (Y) be a maximal circuit-less (cutset-less) subset of a graph and let x^* (y^*) belong to $E \setminus X$ ($E \setminus Y$). If an edge x (y) belongs to the fundamental circuit (cutset) with respect to X (Y), defined by x^* (y^*), then $(X \setminus \{x\}) \cup \{x^*\}$ $((Y \setminus \{y\}) \cup \{y^*\})$ is also a maximal circuit-less (cutset-less) subset of the graph.*

Proof Let an edge x (y) belong to the fundamental circuit (cutset) C_{x^*} (S_{y^*}) defined by $x^* \in E \setminus X$ ($y^* \in E \setminus Y$). Because this fundamental circuit (cutset) is unique, the subset $X' = (X \setminus \{x\}) \cup \{x^*\}$ ($Y' = (Y \setminus \{y\}) \cup \{y\}^*$) is circuit-less (cutset-less). The subsets X (Y) and X' (Y') have the same cardinality. Since X (Y) is a maximal circuit-less (cutset-less) subset, according to Proposition 2.12, X' (Y') is also a maximal circuit-less (cutset-less) subset of the graph. $\square$

The following proposition connects maximal circless sets with maximal cutless sets.

Proposition 2.17 *Let E be the edge set of a graph. Then X (Y) is a maximal circless (cutless) subset of E iff $E \setminus X$ ($E \setminus Y$) is a maximal cutless (circless) subset of E.*

Proof The proof is by contradiction. Let X be a maximal circless subset of E and suppose there is a cut S that entirely belongs to $E \setminus X$. Let e be an edge in S. Then e forms both a circuit with edges in X only and a cutset with edges in $E \setminus X$ only. According to the Orthogonality Theorem, this is a contradiction. In a completely dual manner we can prove that the existence of a circuit in $E \setminus Y$ leads to a contradiction. This completes the proof. $\square$

Corollary 2.18 [to Proposition 2.17] *Let E be the edge set of a graph G. Then* $\operatorname{rank}(E) + \operatorname{corank}(E) = |E|$.

Proof Let X be a maximal cutless subset of E. Then (Proposition 2.17), $E \setminus X$ is a maximal circless subset of E. But rank $(E) = |X|$, corank $(E) = |E \setminus X|$ and $|X| + |E \setminus X| = |E|$, which completes the proof. $\qquad\square$

Remark 2.19 Let E' be a proper nonempty subset of the edge set E of a graph. Then it is not generally true that rank (E') + corank $(E') = |E'|$. For example, for an edge subset E' of a graph that consists of a single edge which forms neither a self-circuit nor a self-cutset, rank $(E') =$ corank $(E') = |E'| = 1$ and hence rank (E') + corank $(E') = 2 \neq 1 = |E'|$.

Corollary 2.20 [to Corollary 2.18 and Propositions 2.7 and 2.17] *For a graph with edge set E, let* $\mathbf{A}$ *be a maximal independent collection of cuts and let* $\mathbf{B}$ *be a maximal independent collection of circs. Then,* $|\mathbf{A}| =$ rank (E), $|\mathbf{B}| =$ corank (E) *and* $|\mathbf{A}| + |\mathbf{B}| = |E|$.

Proof Let X be a maximal circless subset of the edge set E of a graph. Then from Proposition 2.17, $Y = E \setminus X$ is a maximal cutless subset of E. Obviously, $|X| + |E \setminus X| = |E|$. Let $\mathbf{A}_X$ ($\mathbf{B}_Y$) be a maximal independent collection of cuts (circs) in correspondence with the maximal circless (cutless) subset X (Y). According to Proposition 2.7, the cardinality of $\mathbf{A}_X$ ($\mathbf{B}_Y$) is equal to the cardinality of X (Y) and hence $|\mathbf{A}_X| =$ rank (E) ($|\mathbf{B}_Y| =$ corank (E)). But, from Proposition 2.3, all maximal independent collections of cuts (circs) are of the same cardinality and therefore the equality $|\mathbf{A}| =$ rank (E) ($|\mathbf{B}| =$ corank (E)) is valid for any maximal independent collection $\mathbf{A}$ ($\mathbf{B}$) of cuts (circs). On the other hand, from Corollary 2.18, rank (E) + corank $(E) = |E|$ and consequently, $|\mathbf{A}| + |\mathbf{B}| = |E|$ for any maximal independent collection $\mathbf{A}$ of cuts and any maximal independent collection $\mathbf{B}$ of circs. $\qquad\square$

Remark 2.21 We can relate the rank and the corank with other integer parameters of the graph such as the cardinality of the vertex set $|V|$, the cardinality of the edge set $|E|$ and the number of 1-connected components of G. If a graph is connected then in any maximal circuit-less subset of the graph, the number of edges is equal to $|V| - 1$. Hence, the rank of the connected graph is equal to $(|V| - 1)$. According to Corollary 2.18, the corank of a graph is equal to $|E| -$ rank (E). Therefore for a connected graph the corank is equal to $|E| - (|V| - 1)$. Let G be a graph with p components $G_1, \ldots, G_p$, each of which is 1-connected and let $V_1, \ldots, V_p$ and $E_1, \ldots, E_p$, respectively, be the associated vertex and edge sets. Finally, the cardinality of a maximal circless subset of the edge subset E of a graph G is equal to the sum of the cardinalities of a maximal circless subsets of its components $G_k, 1 \leq k \leq p$. Consequently,

$$
\begin{aligned}
\text{rank}\,(E) &= \text{rank}\,(E_1) + \cdots + \text{rank}\,(E_p) \\
&= (|E_1| - (|V_1| - 1)) + \cdots + (|E_p| - (|V_p| - 1)) \\
&= (|E_1| + \cdots + |E_p|) - ((|V_1| + \cdots + |V_p|) - p) \\
&= |E| - (|V| - p)
\end{aligned}
$$

Remark 2.22 As an immediate consequence of Proposition 2.17, the following statement holds: *Let E be the edge set of a graph G and let t be a maximal circuit-less subset of G. Then, $E \setminus t$ is a maximal cutset-less subset of G.*

Using arguments similar to those of Corollary 1.29 of Chapter 1, and taking into account Remark 2.22, it is easy to see that the following statement, holds:

Corollary 2.23 [to the Painting and Orthogonality Theorems] *Given a graph G with edge set E, let t be a maximal circuit-less subset and t^* be an associated maximal cutset-less subset of G. Let e be an edge in t and let e^* be an edge in t^*. Then, e belongs to the circuit that e^* forms with edges in t only iff e^* belongs to a cutset that e forms with edges in t^* only.*

Trees have the collective properties stated in the following remark which is a direct consequence of Proposition 2.16.

Remark 2.24 Let $\mathbf{T}$ be a collection of all trees of a graph. Then, the following conditions hold:

(T1) *No member of $\mathbf{T}$ contains another member as a proper subset.*

(T2) *If $t_1, t_2 \in \mathbf{T}$, and $x_1 \in t_1 \setminus t_2 = t_1 \cap t_2^*$, then there exists $x_2 \in t_2 \setminus t_1 = t_2 \cap t_1^*$ such that $(t_2 \setminus \{x_2\}) \cup \{x_1\}) \in \mathbf{T}$.*

Corollary 2.25 [to the Orthogonality Theorem and Proposition 2.16] The following seemingly *stronger version* of (T2) is also true:

(T2)* *For any two trees t_1 and t_2 of a graph and any $y \in t_2$, the number of edges $x \in t_1$ (not necessarily distinct from y), such that $t_1 \setminus \{x\}) \cup \{y\}$ and $t_2 \setminus \{y\}) \cup \{x\}$ are trees of the graph, is odd.*

Proof If $y \in t_1 \cap t_2$, then the statement is trivially verified, because in that case x and y must be equal. Suppose now $y \notin t_1$ and denote by $C(y, t_1)$ the fundamental circuit formed by y with respect to the maximal circuit-less subset t_1, and by $S(y, t_2)$ the fundamental cutset formed by y relative to the maximal cutset-less subset $E \setminus t_2$. It is easy to see that $(t_1 \setminus \{x\}) \cup \{y\}$ is a maximal circuit-less subset iff $x \in C(y, t_1)$, and similarly that $(t_2 \setminus \{y\}) \cup \{x\}$ is a maximal circuit-less subset iff $x \in S(y, E \setminus t_2)$. Since, according to the Orthogonality Theorem (Proposition 1.20, Chapter 1) the set $C(y, t_1) \cap S(y, t_2)$ has an even number of edges, and moreover has only the edge y not in t_1 (as does $C(y, t_1)$), there are an odd number of edges $x \in t_1 \cap C(y, t_1) \cap S(y, t_2)$. This completes the proof. □

Given a maximal circuit-less subset t, any edge in the corresponding maximal cutset-less subset t^* forms exactly one circuit with edges in t, (called a *fundamental circuit* of G with respect to t). Similarly, any edge of the maximal circuit-less subset t defines exactly one cutset with the edges in the corresponding maximal cutset-less subset t^* (called a *fundamental cutset* of G with respect to t^*).

Lemma 2.26 *Let A be a circuit-less subset of edges of a graph G and let B be a cutset-less subset of G, such that $A \cap B = \emptyset$. Then there exists a maximal circuit-less subset t of G such that $A \subseteq t$ and $B \cap t = \emptyset$.*

Proof Let t_A be a maximal circuit-less subset that contains A and let t_B be a maximal circuit-less subset that does not contain B. Then, $t_B^* \supseteq B$. If $t_B^* \cap A = \emptyset$, then obviously $A \subseteq t_B$ and $B \cap t_B = \emptyset$ and consequently, the proposition is immediately fulfilled for $t = t_B$. Similarly if $t_A \cap B = \emptyset$ then immediately $t = t_A$. Suppose $t_B^* \cap A \neq \emptyset$ and $t_A \cap B \neq \emptyset$. Then the proof is by construction according to the following procedure related to the set t_A: Pick $x \in t_A \cap B$. Because t_A and t_B are trees of G and $x \in t_A$, according to Remark 2.24, there exists $y \in t_A^* \cap t_B$ so that $t_A' = (t_A \setminus \{x\}) \cup \{y\}$ is a maximal circuit-less subset. Because no element of A or B belongs to $t_A^* \cap t_B$, t_A' still contains A and t_B^* contains B. If $t_A' \cap B = \emptyset$ then immediately, $t = t_A'$. Otherwise, if $t_A' \cap B \neq \emptyset$, set $t_A \leftarrow t_A' = (t_A \setminus \{x\}) \cup \{y\}$ and repeat the procedure. After a finite number of repetitions the process terminates when $t_A' \cap B \neq \emptyset$. $\square$

Corollary 2.27 [to Lemma 2.26] *Let E' be a subset of the edge set E of a graph G. Then the following relations hold:*
$$|E'| - \operatorname{corank}(E') = \operatorname{rank}(E) - \operatorname{rank}(E \setminus E'),$$
$$|E'| - \operatorname{rank}(E') = \operatorname{corank}(E) - \operatorname{corank}(E \setminus E').$$

Proof Let B be a maximal cutset-less subset of E' and let A be a maximal circuit-less subset of $E \setminus E'$. Then $|B| = \operatorname{corank}(E')$ and $|A| = \operatorname{rank}(E \setminus E')$. Since $A \cap B = \emptyset$, according to Lemma 2.26, there exists a maximal circuit-less subset t of G so that $A \subseteq t$ and $B \cap t = \emptyset$. Because $t \cap (E \setminus E')$ is a circuit-less subset of $E \setminus E'$, that contains A, and because A is a maximal circuit-less subset of $E \setminus E'$, we conclude that $t \cap (E \setminus E') = A$. Also, because $(E \setminus t) \cap E'$ is a cutset-less subset of E' that contains B, and because B is a maximal cutset-less subset of E' we conclude that $(E \setminus t) \cap E' = B$. Therefore, $t \setminus A = t \cap E' = E' \setminus B$. Obviously, $|B| = |E'| - |E' \setminus B|$ and hence $\operatorname{corank}(E') = |B|$ $= |E'| - |t \setminus A| = |E'| - (|t| - |A|) = |E'| - (\operatorname{rank}(E) - \operatorname{rank}(E \setminus E'))$. Thus, $|E'| - \operatorname{corank}(E') = \operatorname{rank}(E) - \operatorname{rank}(E \setminus E')$. This relation also holds if we replace $E \setminus E'$ by E' and E' by $E \setminus E'$, that is, $|E \setminus E'| - \operatorname{corank}(E \setminus E') = \operatorname{rank}(E) - \operatorname{rank}(E')$. On the other hand, $|E \setminus E'| = |E| - |E'|$ and according to Corollary 2.18, $\operatorname{rank}(E) = |E| - \operatorname{corank}(E)$. Hence, we finally obtain $|E'| - \operatorname{rank}(E') = \operatorname{corank}(E) - \operatorname{corank}(E \setminus E')$. $\square$

Lemma 2.28 *Let t be a maximal circuit-less subset of a graph G and let C be a circuit. Then the following relation holds: $C = \bigoplus_{e \in C \setminus t} C(e,t)$, where $C(e,t)$ is the fundamental circuit that $e \in C \setminus t$ forms with edges in t only.*

Proof Suppose $C \neq \bigoplus_{e \in C \setminus t} C(e,t)$, that is, suppose $(\bigoplus_{y \in C \setminus t} C(y,t)) \oplus C$ is not empty. Because every $y \in C \setminus t$ appears exactly once in $C(y,t)$ and once in C, we conclude that $(\bigoplus_{y \in C \setminus t} C(y,t)) \oplus C \subset t$. But because $(\bigoplus_{y \in C \setminus t} C(y,t)) \oplus C$ is nonempty, it must be a circuit or a union of circuits, which is a contradiction.   □

Dual of Lemma 2.28 *Let t^* be a maximal cutset-less subset of a graph G and let S be a cutset of G. Then the ring sum $S = \bigoplus_{e \in S \setminus t^*} S(e,t^*)$, where $S(e,t^*)$ is the fundamental cutset that $e \in S \setminus t^*$ forms with edges in t^* only.*

Corollary 2.29 [to Lemma 2.28] *Let t be a maximal circuit-less subset of a graph G and let C be a circuit of G such that $t \cap C$ is nonempty. Then for each edge x in $t \cap C$ there exists an edge y in $C \setminus t$ such that x belongs to the fundamental circuit $C(y,t)$ defined by y with respect to t.*

Proof The proof is by contradiction. Suppose there is an edge x in $t \cap C$ such that no fundamental circuit, $C(y,t)$, defined by an edge $y \in C \setminus t$, contains x. Because $x \in C$ and $x \notin C(y,t)$, for any $y \in C \setminus t$, we conclude that $(\bigoplus_{y \in C \setminus t} C(y,t)) \oplus C$ contains x. But, according to Lemma 2.28, $(\bigoplus_{y \in C \setminus t} C(y,t)) \oplus C$ is empty which is a contradiction.   □

Dual of Corollary 2.29: *Let t^* be a maximal cutset-less subset of a graph G and let S be a cutset of G such that $t^* \cap S$ is nonempty. Then for each edge x in $t^* \cap S$ there exists an edge y in $S \setminus t^*$ such that x belongs to the fundamental cutset $S(y,t^*)$ defined by y with respect to t^*.*

Proposition 2.30 *Given an arbitrary pair of trees (t_1, t_2) of a graph the following statements hold:*

(i) Each edge in $t_2 \setminus t_1$ defines a fundamental circuit with respect to t_1 that contains an edge in $t_1 \setminus t_2$.

(ii) Each edge in $t_1 \setminus t_2$ belongs to a fundamental circuit defined by an edge in $t_2 \setminus t_1$ with respect to t_1.

Proof (i) The proof is by contradiction. Suppose there is an edge y in $t_2 \setminus t_1$ such that the fundamental circuit defined by y with respect to t_1, contains no edges in $t_1 \setminus t_2$. Then this fundamental circuit belongs entirely to t_2, a contradiction.

(ii) Let x be an edge in $t_2 \setminus t_1$. From part (i), x defines a fundamental circuit C with respect to t_1 that contains an edge, say y, in $t_1 \setminus t_2$. Since C has a nonempty intersection with t_2 (contains x) we conclude from Corollary 2.29 that there exists an edge z in $C \setminus t_1$ such that y belongs to the fundamental circuit defined by z with respect to t_1. This completes the proof.   □

Proposition 2.31 *Given an arbitrary pair of trees* (t_1, t_2) *of a graph, let an edge* x *belong to* $t_1 \cap t_2$. *If* x *belongs to a fundamental circuit defined by an edge* y_1 *in* $t_1 \setminus t_2$ *with respect to* t_2, *then* x *also belongs to a fundamental circuit defined by an edge* y_2 *in* $t_2 \setminus t_1$, *with respect to* t_1.

Proof Let y_1 be an edge in $t_1 \setminus t_2$ such that the fundamental circuit C defined by y_1 with respect to t_2, contains an edge x in $t_1 \cap t_2$. Since $t_1 \cap C$ is nonempty (contains x), we have from Corollary 2.29 that there exists an edge y_2 in $C \setminus t_1$ such that x belongs to the fundamental circuit defined by y_2 with respect to t_1. ☐

2.4 Circ and cut vector spaces

Let $\mathbf{P}(G)$ be the collection of all edge subsets of a graph G and $\oplus$ be the set operation called ring sum. It is clear that ring sum is closed in $\mathbf{P}(G)$ and therefore the pair $(\mathbf{P}(G), \oplus)$ forms a commutative group. If, in addition, we provide an external product $(\bullet : \mathbf{P}(G) \times \mathsf{F} \to \mathbf{P}(G))$, where F is a suitably chosen field, we may establish a vector structure on $\mathbf{P}(G)$ over this field. From the following properties of ring sum: $Q \oplus \emptyset = Q$, $\emptyset \oplus Q = Q$ and $Q \oplus Q = \emptyset$, for any $Q \in \mathbf{P}(G)$, we conclude that a suitable field is the Galois field $GF(2)$. Recall that $GF(2)$ consists of two elements usually denoted by 1 and 0, and two operations ($+$ and $\cdot$). In this field the following hold:

$0 + 0 = 0$, $1 + 0 = 0 + 1 = 1$, $1 + 1 = 0$,

$0 \cdot 0 = 0$, $1 \cdot 0 = 0 \cdot 1 = 0$, $1 \cdot 1 = 1$.

Let $(\bullet : \mathbf{P}(G) \times F \to \mathbf{P}(G))$ be a map that defines an external product of the elements of $\mathbf{P}(G)$ by the scalars in the Galois field $GF(2)$ as follows:

$\forall A \in \mathbf{P}(G), 1 \bullet \mathbf{A} = \mathbf{A}$ and $0 \bullet \mathbf{A} = \emptyset$.

It is then not difficult to prove that $\forall A, B, C \in \mathbf{P}(G)$ and $\forall \alpha, \beta, \gamma \in GF(2)$ the following list of relations concerning the internal operation $\oplus$ and external operation $\bullet$ hold in $\mathbf{P}(G)$:

$(\alpha + \beta) \bullet C = (\alpha \bullet C) \oplus (\beta \bullet C)$,

$\gamma \bullet (A \oplus B) = (\gamma \bullet A) \oplus (\gamma \bullet B)$,

$(\alpha \cdot \beta) \bullet C = \alpha \bullet (\beta \bullet C)$.

Consequently, the triple $(\mathbf{P}(G), \oplus \mid \bullet)$ is a vector space over the Galois field $GF(2)$. In this vector space the linear combination of the members F_1, ..., F_j of $\mathbf{P}(G)$ is defined to be the expression $(\gamma_1 \bullet F_1) \oplus \cdots \oplus (\gamma_j \bullet F_j)$ where $\gamma_1, \ldots, \gamma_j \in GF(2)$.

Let G be a graph with b edges, $e_1, \ldots, e_b$ and let $Q_k = \{e_k\}$, $k = 1$, ..., b. Obviously, $\mathbf{Q} = \{Q_1, \ldots, Q_b\}$ is a maximal independent collection of members of $\mathbf{P}(G)$ and according to Lemma 2.2, $\mathbf{Q}$ generates $\mathbf{P}(G)$, that is, $\langle \mathbf{Q} \rangle = \{\emptyset, Q_1\} \oplus \cdots \oplus \{\emptyset, Q_m\} = \mathbf{P}(G)$. Hence, according to Lemma 2.2 and Proposition 2.3, any other independent collection of b edge subsets of G generates $\mathbf{P}(G)$.

Accordingly, each member of $\mathbf{P}(G)$ can be represented in a unique way as the ring sum of a number (which is less or equal to b) of edge subsets that belong to a maximal collection of edge subsets. Therefore, any maximal collection of edge subsets $\mathbf{F} = \{F_1, \ldots, F_b\}$ can be seen as a *basis* of the vector space $(\mathbf{P}(G), \oplus \mid \bullet)$, which means that each member of $\mathbf{P}(G)$ can be represented in a unique way as a linear combination of the members of $\mathbf{F} = \{F_1, \ldots, F_b\}$, with coefficients in $GF(2)$.

From Proposition 1.11 of Chapter 1, (respectively, Proposition 1.16 of Chapter 1), the ring sum of any two cuts (circs) is also a cut (circ) and therefore the collection $\mathbf{S}(G)$ of all cuts and the collection $\mathbf{C}(G)$ of all circs are also closed with respect to the ring sum operation. Because $\mathbf{S}(G)$ and $\mathbf{C}(G)$ are subsets of $\mathbf{P}(G)$, it follows that $(\mathbf{S}(G), \oplus)$ and $(\mathbf{C}(G), \oplus)$ are also commutative groups.

Clearly, the collection $\mathbf{C}(G)$ $(\mathbf{S}(G))$ with the internal binary operation $\oplus$ and the external product operation $\bullet$ forms a vector space $(\mathbf{C}(G), \oplus \mid \bullet)$ $((\mathbf{S}(G), \oplus \mid \bullet))$ over the field $GF(2)$. Let $\mathbf{A} = \{C_1, \ldots, C_m\}$ $(\mathbf{B} = \{S_1, \ldots, S_n\})$ be a maximal independent collection of members of $\mathbf{C}(G)$ $(\mathbf{S}(G))$. Then, according to Lemma 2.2, $\mathbf{A}$ $(\mathbf{B})$ generates $\mathbf{C}(G)$ $(\mathbf{S}(G))$, that is, $\langle \mathbf{A} \rangle = \mathbf{C}(G)$ $(\langle \mathbf{B} \rangle = \mathbf{S}(G))$ and $\mathbf{A}$ $(\mathbf{B})$ can be seen as a basis of the vector space $(\mathbf{C}(G), \oplus \mid \bullet)$ $((\mathbf{S}(G), \oplus \mid \bullet))$. That is, each member of $\mathbf{C}(G)$ $((\mathbf{S}(G))$ can be represented in a unique way via the members of $\mathbf{A}$ $(\mathbf{B})$ as a linear combination $(\alpha_1 \bullet C_1) \oplus \cdots \oplus (\alpha_m \bullet C_m)$, where $\alpha_1, \ldots, \alpha_m \in GF(2)$ $((\beta_1 \bullet S_1) \oplus \cdots \oplus (\beta_n \bullet S_n)$, where $\beta_1, \ldots, \beta_n \in GF(2))$.

Using the terminology introduced in this section, we may say:
A graph G is Eulerian iff there exist m numbers $\alpha_1, \ldots, \alpha_m \in GF(2)$ such that $(\alpha_1 \bullet C_1) \oplus \cdots \oplus (\alpha_m \bullet C_m) = E$. Dually, *a graph G is bipartite iff there exist n numbers $\beta_1, \ldots, \beta_n \in GF(2))$ such that $(\beta_1 \bullet S_1) \oplus \cdots \oplus (\beta_n \bullet S_n) = E$.*

Let $\{C_1, \ldots, C_m\}$ be a maximal collection of circs of G and let $\{S_1, \ldots, S_n\}$ be a maximal collection of cuts of G. Notice that, according to Corollary 2.20, $m + n = b$ as is the number of members in any maximal independent collection that generates $\mathbf{P}(G)$. The *hybrid collection* $\mathbf{H} = \{C_1, \ldots, C_m, S_1, \ldots, S_n\}$ of circs and cuts might not be independent. Therefore for an edge subset F of a graph, the question is: does there exist a collection of disjoint circs and cuts, covering F?

Using the above terminology we can preformulate the above question as follows: do there exist $(m + n)$ numbers $\alpha_1, \ldots, \alpha_m, \beta_1, \ldots, \beta_n \in GF(2)$ such that $((\alpha_1 \bullet C_1) \oplus \cdots \oplus (\alpha_m \bullet C_m)) \oplus ((\beta_1 \bullet S_1) \oplus \cdots \oplus (\beta_n \bullet S_n)) = F$. In particular, does there exists such a collection of $(m + n)$ numbers when F coincides with the whole edge set E or when F is the empty set? The answers to these questions are stated in Assertions 1.30 and 1.31 of Chapter 1. We provide equivalent formulations in this section.

Remark 2.32 Suppose there exist $(m+n)$ numbers $\alpha_1, \ldots, \alpha_m, \beta_1, \ldots, \beta_n \in$

$GF(2)$ such that $((\alpha_1 \bullet C_1) \oplus \cdots \oplus (\alpha_m \bullet C_m)) \oplus ((\beta_1 \bullet S_1) \oplus \cdots \oplus (\beta_n \bullet S_n)) = \emptyset$. Since $(\alpha_1 \bullet C_1) \oplus \cdots \oplus (\alpha_m \bullet C_m)$ is a nonempty circ C and $(\beta_1 \bullet S_1) \oplus \cdots \oplus (\beta_n \bullet S_n)$ is a nonempty cut S. From $C \oplus S = \emptyset$ it follows that $C = S$. Thus there is a circ which is at the same time a cut.

In order to make an explicit connection between the vector structures introduced above with standard linear algebra, we introduce a special representation for edge subsets of a graph. For that reason, we order the edge set E in an arbitrary way by introducing a map $f : E \rightarrow \{1, \ldots, b\}$. Then, consider b-tuples $(k_1, \ldots, k_b)$ where $k_i \in \{0,1\}$. Obviously, to each edge subset F of G we can associate a b-tuple $(k_1, \ldots, k_b)$ where for $1 \leq j \leq b$, $k_j = 1$ if e_j belongs to F, and $k_j = 0$ if e_j does not belong to F. The converse is also true: to each b-tuple $(k_1, \ldots, k_b)$ where $k_i \in \{0,1\}$ we can associate an edge subset F so that for $1 \leq j \leq b$, e_j belongs to F if $k_j = 1$, and e_j does not belong to F if $k_j = 0$. Since $\mathbf{P}(G)$ is the collection of all edge subsets of a graph G, there is a bijection between $\mathbf{P}(G)$ and the set $\mathcal{P}(G)$ of all b-tuples $(k_1, \ldots, k_b)$ with $k_i \in \{0,1\}$. From $\langle \mathbf{Q} \rangle = \{\emptyset, Q_1\} \oplus \cdots \oplus \{\emptyset, Q_m\} = \mathbf{P}(G)$ it follows that $\mathcal{P}(G)$ can be represented as the Cartesian product of b identical sets $\{0,1\}$, that is, $\mathcal{P}(G) = \{0,1\} \times \cdots \times \{0,1\}$.

Let us now introduce an internal operation $+$ in the set $\mathcal{P}(G)$ defined through the operation $+$ of the Galois field as follows:

(O1) $(k_1, \ldots, k_b) + (l_1, \ldots, l_b) = (k_1 + l_1, \ldots, k_b + l_b)$,

and an external operation $\circ : \mathcal{P}(G) \times \{0,1\} \rightarrow \mathcal{P}(G)$ defined through the operations $\cdot$ of the Galois field as follows:

(O2) $\alpha \circ (k_1, \ldots, k_b) = (\alpha \cdot k_1, \ldots, \alpha \cdot k_b)$,

where $(k_1, \ldots, k_b), (l_1, \ldots, l_b) \in \mathcal{P}(G)$ and $\alpha \in \{0,1\}$.

It is easy to see that $(\mathcal{P}(G), + \mid \circ)$ is a vector space isomorphic to the vector space $(\mathbf{P}(G), \oplus \mid \bullet)$. Therefore, from now on, we shall interpret b-tuples of 1's and 0's as $b \times 1$ column vectors of the vector space $(\mathcal{P}(G), + \mid \circ)$.

Denote by $(\mathcal{C}(G), + \mid \circ)$ and $(\mathcal{S}(G), + \mid \circ)$ the vector subspaces of the vector space $(\mathcal{P}(G), + \mid \circ)$ that are induced by the vector subspaces $(\mathbf{C}(G), \oplus \mid \bullet)$ and $(\mathbf{S}(G), \oplus \mid \bullet)$.

According to the above representation, with every edge subset of G we have an associated $b \times 1$ column vector of 1's and 0's. Consequently, every q-tuple of subsets of a graph can be interpreted in $GF(2)$ as a $b \times q$ matrix in which every column corresponds to exactly one subset in the q-tuple. The members of the matrix are 1's and 0's, that is the elements of $GF(2)$. In particular, every b-tuple of members of a hybrid collection of circs and cuts can be interpreted as a $b \times b$ matrix, in which every column corresponds to a circ or a cut. In what follows we shall use a special hybrid collection of circs and cuts that consists of fundamental cutsets and fundamental circuits with respect to a maximal circuit-less subset of a graph. Let $(e_1, \ldots, e_n, e_{n+1}, \ldots, e_{n+m})$ be an $(n + m)$-tuple of edges of G such that the first n places correspond to the edges that belong to a maximal circuit-less subset of G and the last m to the edges that belong to a maximal cutset-less subset of G. Each edge of G, with respect to a maximal circuit-less subset t of G, is associated in a unique manner either with a fundamental cutset or with a fundamental circuit with respect to a maximal circuit-less subset. Therefore such an $(n + m)$-tuple of edges is associated with the following $(n + m)$-tuple of fundamental cutsets and circuits $(S_1, \ldots, S_n, C_1, \ldots, C_m)$. Clearly, $\{S_1, \ldots, S_n, C_1, \ldots, C_m\}$ is a hybrid collection of circs and cuts.

According to our representation of edge subsets as $b \times 1$ columns of 1's and 0's, every fundamental cutset S and every fundamental circuit C of the hybrid collection can be represented as a column vector whose entries are 1's and 0's. Suppose that the members of each column are placed in the same order, from top to bottom as the members in the $(n+m)$-tuple $(e_1, \ldots, e_n, e_{n+1}, \ldots, e_{n+m})$ from left to right. Accordingly, in the matrix $\mathsf{H}^{\mathrm{T}} = (\mathsf{S}_1, \ldots, \mathsf{S}_n, \mathsf{C}_1, \ldots, \mathsf{C}_m)$ both columns and rows are ordered in the same manner. Because $S_1, \ldots, S_n$, are fundamental cutsets with respect to a maximal circuit-less subset t and $C_1, \ldots, C_m$ are fundamental circuits with respect to the associated maximal cutset-less subset t^*, and taking into account the fact that an edge $e \in t$ belongs to the fundamental circuit that an edge $e^* \in t^*$ forms with edges in t only iff e^* belongs to the fundamental cutset that e forms with edges in t^* only (Corollary 2.23), we conclude that the matrix H is symmetric, that is, $\mathsf{H} = \mathsf{H}^{\mathrm{T}}$. Consequently, if H_k is the k-th column of H then $\mathsf{H}_k^{\mathrm{T}}$ is the k-th row of H. This can be also proved by using well-known properties of the matrices $\mathsf{Q}^{\mathrm{T}} = (\mathsf{S}_1, \ldots, \mathsf{S}_n)$, and $\mathsf{B}^{\mathrm{T}} = (\mathsf{C}_1, \ldots, \mathsf{C}_m)$, called the *fundamental cutset matrix* and the *fundamental circuit matrix*, respectively and the fact that $\mathsf{H} = [\mathsf{Q}; \mathsf{B}]$. From now on, we shall call the matrix H a *fundamental hybrid matrix*.

The following proposition is an equivalent formulation of Assertion 1.30 of Chapter 1.

Proposition 2.33 *Let* $\mathsf{H} = \{S_1, \ldots, S_n, C_1, \ldots, C_m\}$ *be a hybrid collection of circs and cuts of a graph* G *Then, for each edge subset* F *of* G *there exist* $(m + n)$ *numbers* $\alpha_1, \ldots, \alpha_m, \beta_1, \ldots, \beta_n \in GF(2)$ *such that* $(\alpha_1 \bullet C_1) \oplus \cdots \oplus (\alpha_m \bullet C_m) \oplus (\beta_1 \bullet S_1) \oplus \cdots \oplus (\beta_n \bullet S_n) = F$, *iff there is no circ of* G *which is at the same time a cut of* G.

Proof Let $\mathsf{H}^{\mathrm{T}} = (\mathsf{S}_1, \ldots, \mathsf{S}_n, \mathsf{C}_1, \ldots, \mathsf{C}_m)$ be the fundamental hybrid matrix associated with the hybrid collection H of a graph G. According to the definition of the fundamental hybrid matrix H, it is a symmetric matrix in which columns and the rows are taken in the same order. In the matrix H, the rows formally correspond to fundamental cutsets and fundamental circuits of G and columns to edges. Then using the sequence of equivalent transformations (well-known in linear algebra) that preserve rank, we can transform the matrix H into another L, in which all members L_{ij}, $j \leq i$

are equal to zero. This type of reduction is called a reduction to *row echelon matrix form*. Notice that rows of the matrix L correspond to edge subsets that can be represented as ring sums of fundamental cutsets and circuits. If there is no row in the matrix L corresponding to an empty set then L has maximal rank equal to b, and consequently, H also has rank equal to b. This means that the collection of edge subsets that correspond to the collection of rows of H is a maximal independent collection in the set $\mathbf{P}(G)$ and therefore generates $\mathbf{P}(G)$. Otherwise, if there are rows that correspond to empty sets, the initial hybrid collection of circs and cuts is not independent and therefore does not generate the set $\mathbf{P}(G)$. If there is a subcollection of a hybrid collection of circs and cuts of a graph whose ring sum is the empty set, according to Remark 2.32, we conclude that there is a circ of the graph which is at the same time a cut of the graph. $\qquad\square$

The following proposition is an equivalent formulation of Assertion 1.31 of Chapter 1.

Proposition 2.34 *Let G be a graph with edge set E and let t be a maximal circuit-less subset of G. Let $\{C_1, \ldots, C_m\}$ be the collection of all fundamental circuits with respect to t and let $\{S_1, \ldots, S_n\}$ be the collection of all fundamental cutsets of G, with respect to t. Then, there exist $(m+n)$ numbers $\alpha_1, \ldots, \alpha_m, \beta_1, \ldots, \beta_n \in GF(2)$ such that $(\alpha_1 \bullet C_1) \oplus \cdots \oplus (\alpha_m \bullet C_m) \oplus (\beta_1 \bullet S_1) \oplus \cdots \oplus (\beta_n \bullet S_n) = E$.*

Proof In $GF(2)$, the relation $(\alpha_1 \bullet C_1) \oplus \cdots \oplus (\alpha_m \bullet C_m) \oplus (\beta_1 \bullet S_1) \oplus \cdots \oplus (\beta_n \bullet S_n) = E$ can be interpreted as the following system of linear equations:

$$\mathsf{H}^\mathsf{T}\mathsf{x} = \mathsf{E} \qquad (1)$$

where $\mathsf{x} = (\alpha_1, \ldots, \alpha_m, \beta_1, \ldots, \beta_n)^\mathsf{T}$, $\alpha_1, \ldots, \alpha_m, \beta_1, \ldots, \beta_n \in GF(2)$ and E stands for the column vector consisting of all 1's. Because the matrix H is symmetric, system (1) is equivalent to the system:

$$\mathsf{x}^\mathsf{T}\mathsf{H}^\mathsf{T} = \mathsf{E}^\mathsf{T}. \qquad (1a)$$

Notice that in the matrix H^T, columns correspond to cutsets and circuits and therefore, the system (1a) can be seen as a matrix representation of the following system of scalar equations:

$$\mathsf{x}^\mathsf{T} \circ \mathsf{S}_k = 1, \quad k = 1, \ldots, n \qquad (1s)$$

and

$$\mathsf{x}^\mathsf{T} \circ \mathsf{C}_j = 1, \quad j = 1, \ldots, m. \qquad (1c)$$

The system (1s,1c) has a solution in $GF(2)$ iff we cannot deduce a contradictory equation by forming a linear combination, using the operation from $GF(2)$. Let

$$a_1 \circ \mathsf{x}^{\mathsf{T}} \circ \mathsf{S}_1 + \cdots + a_n \circ \mathsf{x}^{\mathsf{T}} \circ \mathsf{S}_n + b_1 \circ \mathsf{x}^{\mathsf{T}} \circ \mathsf{C}_1 + \cdots + b_m \circ \mathsf{x}^{\mathsf{T}} \circ \mathsf{C}_m$$

$$= a_1 + \cdots + a_n + b_1 + \cdots + b_m \tag{2}$$

be an arbitrary linear combination of a number of the above equations, with coefficients in $GF(2)$. Then the system (1s,1c) has a solution in $GF(2)$ iff the following implication holds: if the left side of (2) is identically equal to zero, then the right side of (2) is also equal to zero, that is, the system (1s,1c) has a solution in $GF(2)$ iff

$$\mathsf{x}^{\mathsf{T}} \circ (a_1 \circ \mathsf{S}_1 + \cdots + a_n \circ \mathsf{S}_n + b_1 \circ \mathsf{C}_1 + \cdots + b_m \circ \mathsf{C}_m) \equiv 0 \tag{2left}$$

implies

$$a_1 + \cdots + a_n + b_1 + \cdots + b_m = 0. \tag{2right}$$

The identity (2 left) holds for every x, iff the relation

$$(a_1 \bullet C_1) \oplus \cdots \oplus (a_m \bullet C_m) \oplus (b_1 \bullet S_1) \oplus \cdots \oplus (b_n \bullet S_n) = 0 \tag{3}$$

holds. Therefore to prove that the system (1s,1c) has a solution, it is enough to prove that (3) implies (2 right).

Because $0 + 0 = 1 + 1 = 0$ in $GF(2)$, (3) holds iff

$$(a_1 \bullet C_1) \oplus \cdots \oplus (a_m \bullet C_m) = (b_1 \bullet S_1) \oplus \cdots \oplus (b_n \bullet S_n) \quad (= W) \tag{4}$$

holds, that is, the set W of edges of G is both a cut and a circ. According to the Orthogonality Theorem such an edge subset, which can be seen as an intersection of a cut and a circ, has an even number of elements. On the other hand, the set of edges of G that correspond to 1's of the $(m + n) \times 1$ column $(a_1, \ldots, a_m, b_1, \ldots, b_n)^{\mathsf{T}}$, coincides with the set W where 1's corresponding to the a_k belong to the maximal circuit-less subset t and 1's corresponding to the b_j belong to the maximal cutset-less subset t^*. Since the number of elements in W is even, the number of 1's is also even. But the sum of an even number of 1's is always equal to zero in $GF(2)$, and hence we conclude that (2 right) implies (2 left). This terminates the proof. $\qquad\square$

As an illustration, consider the complete graph G with four vertices, as Figure 1.8 (Chapter 1) shows. With respect to the maximal circuit-less subset $t = \{b, c, e\}$, the edge subsets $S_b = \{b, a, f\}, S_c = \{c, a, d, f\}$, and $S_e = \{e, d, f\}$ are fundamental cutsets and $C_a = \{a, b, c\}, C_d = \{d, c, e\}$ and $C_f = \{f, b, c, e\}$ are fundamental circuits of G. In the associated matrix H, the first three rows correspond to fundamental cutsets of G while the last three rows corresponds to fundamental circuits of G, with respect to the maximal circuit-less subset t. The matrix H has block form with four 3×3 blocks H_{ij} such that diagonal blocks correspond to unit matrices. After transformation of H into row echelon form, we obtain a new matrix L. For the matrix L we consider another block representation into four blocks denoted by L_{ij}, where L_{11} is the 4×4 unit matrix, L_{12} is a 4×2 matrix with an odd number of nonzero elements, L_{21} is a 2×4 zero matrix and L_{22} is the 2×2 zero matrix. With each row of the matrix L we have associated an edge subset which is a ring sum of a number of cutsets and a number of circuits, providing that the last two rows correspond to empty sets. Every nonempty edge subset set that belongs to the collection of edge subsets corresponding to rows of L contains an edge that does not appear in any other member of the collection. Each of the rest of the edges (d and f) appears in an odd number of nonempty subsets of the collection. Consequently, the ring sum of all edge subsets corresponding to rows of L is the set that coincides with the edge set $E = \{a, b, c, d, e, f\}$. Notice that ring sums that correspond to zero rows are of the form (3) and can be therefore represented in the form (4). Consequently, they define circs that are at the same time cuts. There are two such sums that correspond to zero rows: $C_a \oplus C_d \oplus S_b \oplus S_e = 0$ and $C_f \oplus S_b \oplus S_c \oplus S_e = 0$. They define set edges: $C_a \oplus C_d = S_b \oplus S_e = \{a, b, d, e\}$ and $C_f = S_b \oplus S_c \oplus S_e = \{b, c, e, f\}$.

We also introduce a scalar product in $\mathcal{P}(G)$ as a map $\mathcal{P}(G)^2 \to \{0,1\}$ defined as follows:

$$(k_1, \ldots, k_b)(l_1, \ldots, l_b) = k_1 \cdot l_1 + \cdots + k_b \cdot l_b, \qquad (O3)$$

where $+, \cdot$ are the operations in the Galois field and $k_j, l_j \in \{0,1\}$, $1 \le j \le b$.

Moreover, $(\mathcal{P}(G) = \{0,1\}^b, + \mid \circ))$ is also a vector space with $(\mathcal{C}(G), + \mid \circ)$ and $(\mathcal{S}(G), + \mid \circ)$ as its vector subspaces.

As a consequence of the Orthogonality Theorem (see Proposition 1.20 of Chapter 1) we have the following:

Proposition 2.35 *The vector spaces $(\mathcal{C}(G), + \mid \circ)$ and $(\mathcal{S}(G), + \mid \circ)$ are orthogonal in the sense that the scalar product of an arbitrary element in $\mathcal{C}(G)$ and $\mathcal{S}(G)$ is equal to 0.*

Proof Let $(k_1, \ldots, k_b) \in \mathcal{C}(G)$ and let $(l_1, \ldots, l_b) \in \mathcal{S}(G)$. According to the Orthogonality Theorem, the associated circ $C \in \mathbf{C}(G)$ corresponding to $(k_1, \ldots, k_b)$ and the cut $S \in \mathbf{S}(G)$ corresponding to $(l_1, \ldots, l_b)$ have an even number of common edges. Hence, the scalar product (O3) leads to the sum $k_1 \cdot l_1 + \cdots + k_b \cdot l_b$ in which an even number of terms are equal to 1 and the remaining terms are equal to 0. But a sum of even numbers of 1's in the Galois field $GF(2)$ equals 0 which completes the proof. $\square$

Let $\mathbf{A} = \{S_1, \ldots, S_n\}$ ($\mathbf{B} = \{C_1, \ldots, C_m\}$), be a maximal independent collection of members of $\mathbf{S}(G)$ ($\mathbf{C}(G)$). Then, from Lemma 2.1, $\mathbf{A}$ ($\mathbf{B}$) generates $\mathbf{S}(G)$ ($\mathbf{C}(G)$), that is (using the notation of Section 2.1), $\langle \mathbf{A} \rangle = \mathbf{S}(G)$

($\langle \mathbf{B} \rangle = \mathbf{C}(G)$). The subset $\mathcal{A}$ ($\mathcal{B}$) of $\mathcal{P}(G)$, that corresponds to $\mathbf{A}$ ($\mathbf{B}$), therefore forms a base of the vector space $(\mathcal{S}(G), + \mid \circ)$ $((\mathcal{C}(G), + \mid \circ))$. The cardinality of $\mathcal{A}$ ($\mathcal{B}$), which obviously equals the cardinality of $\mathbf{A}$ ($\mathbf{B}$), is called the *dimension* of the vector space $(\mathcal{S}(G), + \mid \circ)$ $((\mathcal{C}(G), + \mid \circ))$. Denote by $\dim \mathcal{S}(G)$ the dimension of the vector space $\mathcal{S}(G)$ and denote by $\dim \mathcal{C}(G)$ the dimension of the vector space $\mathcal{C}(G)$.

The following proposition relates the dimension of the vector space $\mathcal{S}(G)$ and the dimension of the vector space $\mathcal{C}(G)$ with the number of edges of G.

Proposition 2.36 *Let E be the edge set of a graph G. Then,* $\dim \mathcal{S}(G) = \text{rank}\,(E)$, $\dim \mathcal{C}(G) = \text{corank}\,(E)$ *and* $\dim \mathcal{C}(G) + \dim \mathcal{S}(G) = |E|$.

Proof From Corollary 2.20, $|\mathbf{A}| = \text{rank}\,(E)$ and $|\mathbf{B}| = \text{corank}\,(E)$ and $|\mathbf{A}| + |\mathbf{B}| = |E|$, where $\mathbf{A}$ ($\mathbf{B}$) is a maximal independent collection of cuts (circs). Let $\mathcal{A}$ ($\mathcal{B}$) be the subset of $\mathcal{P}(G)$, that corresponds to $\mathbf{A}$ ($\mathbf{B}$). Then $|\mathbf{A}| = |\mathcal{A}| = \dim \mathcal{S}(G)$ and $|\mathbf{B}| = |\mathcal{B}| = \dim \mathcal{C}(G)$, which completes the proof. $\square$

2.5 Binary graphoids and their representations

As we mentioned at the beginning of this chapter, the following two statements hold for graphs:

(i) $(\mathbf{C}, \otimes \mid E)$ and $(\mathbf{S}, \otimes \mid E)$ are commutative groups,

(ii) if $C \in \mathbf{C}$ and $S \in \mathbf{S}$, then $|C \cap S|$ is even.

Clearly, every edge of a graph is incident to at most two vertices and, according to Lemma 1.10 of Chapter 1, any cut can be obtained as a ring sum of a collection of stars. Thus, the following two additional statements also hold:

(iii) there is a subcollection $\mathbf{S}^*$ of $\mathbf{S}$ such that $\mathbf{S}^*$ generates $\mathbf{S}$,

(iv) each element of E belongs to at most two members of $\mathbf{S}^*$.

Conversely, let E be a finite nonempty set and let $\mathbf{C}$ and $\mathbf{S}$ be two collections of subsets of E satisfying (i) and (ii) that is, let $(E, \mathbf{C}, \mathbf{S})$ be a binary graphoid. Let $\mathbf{S}^*$ be a subcollection of $\mathbf{S}$ that satisfies (iii). Such a subcollection of $\mathbf{S}$ we shall call a *cut basis*. It is clear from Lemma 2.1 and Lemma 2.2 that cut bases always exist. It is also clear from Proposition 2.3 that they all have the same cardinality. The cardinality of a cut basis of a binary graphoid is called the *rank of the binary graphoid*. Notice that there may be several cut bases associated with a cut space $\mathbf{S}$. Let S be the ring sum of all members of a cut basis $\mathbf{S}^*$ (it is clear that S belongs to $\mathbf{S}$). Then the set $\mathbf{S}^* \cup \{S\}$ in an *extended cut basis*. It is also clear that extended cut bases always exist.

Remark 2.37 *Every element of E appears in an even number of members of $\mathbf{S}^*$.*

This is an immediate consequence of the fact that S is obtained as the ring sum of all members of a cut basis $\mathbf{S}^*$ and therefore contains only those elements of E which appear in an odd number of members of $\mathbf{S}^*$. extended set $\mathbf{S}^* \cup \{S\}$, which we call an *extended 2-complete cut basis*, also exists. Clearly, each element of E belongs to *exactly* two members of an extended 2-complete cut basis $\mathbf{S}^* \cup \{S\}$ and hence the members of $\mathbf{S}^* \cup \{S\}$ can be interpreted as vertices of a graph. Therefore each 2-complete cut basis defines a graph. A binary graphoid with at least one 2-complete cut basis is called a *graphic binary graphoid*. The class of all graphs that corresponds to a graphic binary graphoid $\mathcal{G} = (E, \mathbf{C}, \mathbf{S})$ coincides with a class of 2-isomorphic graphs.

As an illustration, consider the triple $\mathcal{G}_1 = (E, \mathbf{C}, \mathbf{S})$ where:

$$E = \{e_1, e_2, e_3, e_4, e_5, e_6, e_7\},$$

$\mathbf{S}$ contains edge subsets $\{e_1, e_2\}$, $\{e_1, e_3\}$, $\{e_2, e_3\}$, $\{e_1, e_4, e_5\}$, $\{e_1, e_6, e_7\}$, $\{e_2, e_4, e_5\}$, $\{e_2, e_6, e_7\}$, $\{e_3, e_4, e_5\}$, $\{e_3, e_6, e_7\}$, $\{e_4, e_5, e_6, e_7\}$ including all their unions and $\mathbf{C}$ contains subsets $\{e_1, e_2, e_3, e_4, e_6\}$, $\{e_4, e_5\}$, $\{e_6, e_7\}$, $\{e_1, e_2, e_3, e_5, e_6\}$, $\{e_1, e_2, e_3, e_4, e_7\}$, $\{e_1, e_2, e_3, e_5, e_7\}$ including all their unions. It is obvious that $\mathcal{G}_1$ satisfies conditions (i) and (ii) and consequently defines a binary graphoid. It is easy to see that the subcollections $\mathbf{S}_1^* = \{\{e_1, e_2\}, \{e_2, e_3\}, \{e_1, e_4, e_5\}, \{e_3, e_6, e_7\}\}$, and $\mathbf{S}_2^* = \{\{e_2, e_3\}, \{e_2, e_4, e_5\}, \{e_1 e_6, e_7\}, \{e_3, e_6, e_7\}\}$ of $\mathbf{S}$ satisfy both conditions (iii) and (iv). The ring sum of all members of the subcollection $\mathbf{S}_1^*$ is the set $S_1 = \{e_4, e_5, e_6, e_7\}$ and the ring sum of all members of the subcollection $\mathbf{S}_2^*$ is the set $S_2 = \{e_1, e_4, e_5\}$. The corresponding extended cut bases $\mathbf{S}_1^* \cup \{S_1\}$ and $\mathbf{S}_2^* \cup \{S_2\}$ define two graph structures as illustrated in Figure 2.2. Notice that the structures of Figure 2.2 coincide with the graphs of Figure 2.1.

Among all cut bases of a graphic binary graphoid, besides those that are 2-complete, there might be also be some that are not 2-complete. For example, for the graphic binary graphoid $\mathcal{G}_1$ of the previous example the subcollection $\mathbf{S}_3^* = \{\{e_1, e_2\}, \{e_2, e_3\}, \{e_2, e_4, e_5\}, \{e_2, e_6, e_7\}\}$ is a cut basis but not 2-complete since the element e_2 belongs to four members of $\mathbf{S}^*$. Observe that the ring sum of all members of $\mathbf{S}_3^*$ is the subset $S_3 = \{e_1, e_3, e_4, e_5, e_6, e_7\}$. If we interpret the members of $\mathbf{S}_3^* \cup \{S\}$ as vertices, the corresponding representation is not a graph but a hypergraph as shown in Figure 2.3.

In this hypergraph all edges are incident to exactly two vertices except the edge e_2 which is incident to four vertices. Notice that in the case of binary

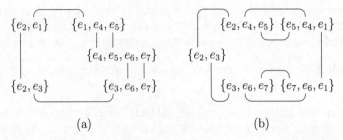

(a)   (b)

Figure 2.2. Two graph structures corresponding to a pair of distinct 2-complete cut bases of the binary graphoid $\mathcal{G}_1$.

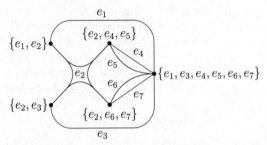

Figure 2.3. A hypergraph corresponding to a non 2-complete cut basis of the graphic binary graphoid $\mathcal{G}_1$.

graphoids we differentiate between *having a representation via hypergraphs* and *being graphic* (which means having a representation via graphs only).

Generally speaking together with every extended cut basis $\mathbf{S}^* \cup \{S\}$ of a binary graphoid $(E, \mathbf{C}, \mathbf{S})$ we can associate a hypergraph. Since every element of E appears in an even number of members of $\mathbf{S}^*$ it follows that each edge of the hypergraph is incident to an even nonzero number of vertices (this includes graphs as a special case). The converse statement is generally not true as the following remark indicates.

Remark 2.38 *A hypergraph, each edge of which is incident to an even nonzero number of vertices, might not be a representation of any triple $(E, \mathbf{C}, \mathbf{S})$ satisfying conditions* (i) *and* (ii).

As a counterexample see the hypergraph of Figure 2.4. The edge subset $\{a, b\}$ of the hypergraph of Figure 2.4 appears twice as a star of this hypergraph which contradicts the obvious fact that members of an extended cut basis must be distinct (the same arguments holds for the edge subset $\{a, c\}$).

There are some binary graphoids with no 2-complete cut bases. Such binary graphoids are called *nongraphic*. As an illustration of a nongraphic binary graphoids consider the triple $\mathcal{G}_2 = (E, \mathbf{C}, \mathbf{S})$, defined on the set $E =$

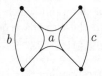

Figure 2.4. A hypergraph which do not correspond to a cut basis of any binary graphoid.

$\{a,b,c,d,e,f,g,h\}$ such that $\mathbf{C}$ contains subsets $\{a,b\}$, $\{b,c,d,e\}$, $\{a,c,d,e\}$, $\{b,c,f,g\}$, $\{a,c,f,g\}$, $\{d,e,f,g\}$, $\{b,d,g,h\}$, $\{a,d,g,h\}$, $\{c,e,g,h\}$, $\{c,d,f,h\}$, $\{b,e,f,h\}$, $\{a,e,f,h\}$, and all their unions and $\mathbf{S}$ contains edge subsets $\{d,e,h\}$, $\{f,g,h\}$, $\{c,e,f\}$, $\{c,d,g\}$, $\{a,b,e,g\}$, $\{a,b,d,f\}$, $\{a,b,c,h\}$, $\{d,e,f,g\}$, $\{c,e,g,h\}$, $\{c,d,f,h\}$, $\{a,b,e,f,h\}$, $\{a,b,d,g,h\}$, $\{a,b,c,f,g\}$, $\{a,b,c,d,e\}$, also including all their unions. It is easy to see that $\mathcal{G}_2$ satisfies conditions (i) and (ii) and therefore is a binary graphoid. Moreover, it is a nongraphic binary graphoid. For example the subcollection $\mathbf{S}_I^* = \{\{a,b,c,h\},$ $\{c,e,f\}$, $\{d,e,h\}$, $\{f,g,h\}\}$ satisfies (iii) but does not satisfy (iv) since the element h belongs to three members of $\mathbf{S}_I^*$. Observe that the ring sum of all members of $\mathbf{S}_I^*$ is the subset $S_I = \{a,b,c,h\}$. If we interpret the members of $\mathbf{S}_I^* \cup \{S_I\}$ as vertices, the corresponding representation is not a graph but a hypergraph shown in Figure 2.5.

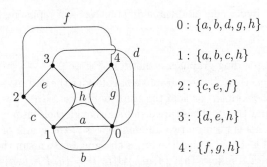

$0 : \{a,b,d,g,h\}$

$1 : \{a,b,c,h\}$

$2 : \{c,e,f\}$

$3 : \{d,e,h\}$

$4 : \{f,g,h\}$

Figure 2.5. A hypergraph corresponding to a cut basis of the nongraphic binary graphoid $\mathcal{G}_2$.

In this hypergraph every edge, except h, is incident to a pair of distinct vertices; the edge h is incident to four vertices. As another example consider the hypergraph that corresponds to the extended basis $\mathbf{S}_{II}^* \cup \{S_{II}\}$ where $\mathbf{S}_{II}^* = \{\{a,b,e,g\}, \{c,e,g,h\}, \{d,e,h\}, \{f,g,h\}\}$ and $S_{II} = \{a,b,c,d,e,f,g,h\}$. It can be seen by inspection that in this hypergraph there are three edges incident to four vertices: e, h and g.

An interesting problem is how to obtain the circ space of a binary graphoid from its cut space using a hypergraph which corresponds to a cut basis. This circ space coincides with the circ space of a *graph* obtained from the hypergraph by substituting all hyper edges with an appropriate collection of regular (graph) edges, as we now describe. Given a binary matroid, let $(E, V; f)$ be a hypergraph corresponding to a cut basis of a binary graphoid $\mathcal{G} = (E, \mathbf{C}, \mathbf{S})$. Then every edge e is incident with an even nonzero number of vertices (if $|f(e)| > 2$ we call it a hyper edge). Consider an edge e of the hypergraph $(E, V; f)$ incident with $2k$ $(k > 1)$ vertices $(|f(e)| = 2k))$ and substitute it with k regular edges (that is, edges incident with exactly two vertices) connecting pairs of vertices from $f(e)$ providing that no two such edges are sharing the same vertex from $f(e)$. Label each of these k regular edges by e. If we repeat this operation until all hyper edges are substituted, we finally obtain a graph with some repeated labels corresponding to the hypergraph. For example the edge h of the hypergraph of Figure 2.5 can be substituted in three different ways by a pair of regular identically labelled edges as it is shown in Figure 2.6.

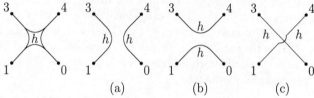

Figure 2.6. A hyper edge and three different ways of replacing it by a pair of regular edges.

The corresponding graphs are shown in Figure 2.7.

The circ spaces of all these graphs coincide with the circ space of the binary graphoid. For example, consider the graph of Figure 2.7(a). Two of its circuits are $\{a, d, h\}$ and $\{h, g\}$. These are disjoint although their intersection contains the edge h. This is because two different edges of the graph of Figure 2.7(a) have the label h. Since we use the terms circ and cut to mean the sets, that is, the subsets, of E it follows that $\{a, d, h\} \cup \{h, g\} = \{a, d, h, h, g\} = \{a, d, h, g\}$. Notice that $\{a, d, h, g\}$ is a circuit of the hypergraph of Figure 2.5. Similarly, $\{a, d, h\}$ and $\{b, c, h, f\}$ are two circuits of the graph of Figure 2.7(a) and their union $\{a, d, h\} \cup \{b, c, h, f\} = \{a, d, h, b, c, h, f\} = \{a, b\} \cup \{c, d, h, f\}$ is the union of two circuits $\{a, b\}$ and $\{c, d, h, f\}$ of the hypergraph of Figure 2.5.

There may be several graphs and/or hypergraphs associated with the same binary graphoid, each with a particular cut base. For the total class of graphs and hypergraphs representing the same binary graphoid we say that they form a class of mutually 2-isomorphic representations.

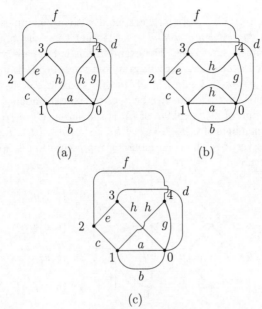

Figure 2.7. Three graphs corresponding to the hypergraph of Figure 2.5.

Every binary graphoid $\mathcal{G} = (E, \mathbf{C}, \mathbf{S})$ has a dual $\mathcal{G}^* = (E, \mathbf{C}^*, \mathbf{S}^*)$, where $\mathbf{C}^* = \mathbf{S}$ and $\mathbf{S}^* = \mathbf{S}$, called a *dual binary graphoid*. Notice that $(\mathcal{G}^*)^* = \mathcal{G}$. The dual of a binary graphoid is obviously also characterized with the same pair of conditions (i) and (ii). Since $\mathbf{C}$ and $\mathbf{S}$ are interchanged, conditions (iii) and (iv) have to be replaced by their duals:

(iii)′ there is a subcollection $\mathbf{C}^*$ of $\mathbf{C}$ such that $\mathbf{C}^*$ generates $\mathbf{C}$

(iv)′ each element of E belongs to at most two members of $\mathbf{C}^*$.

A subcollection $\mathbf{C}^*$ of $\mathbf{C}$ that satisfies (iii)′ is called a *circ basis*. Notice that there may be several circ bases. Let C be the ring sum of all members of a circ basis $\mathbf{C}^*$ (it is clear that C belongs to $\mathbf{C}$). Then the set $\mathbf{C}^* \cup \{C\}$ we call an *extended circ basis*. It is clear from Lemma 2.1 and Lemma 2.2 that circ bases (and extended circ bases) always exist. It is also clear from Proposition 2.3 that they all have the same cardinality. The cardinality of a circ basis of a binary graphoid is called the *corank of the binary graphoid*. It can be shown that the sum of the rank and the corank of a binary graphoid is equal to the cardinality of its ground set. A subcollection $\mathbf{C}^*$ of $\mathbf{C}$ that satisfies both (iii)′ and (iv)′ is called a *2-complete circ basis*. The associated extended cut basis $\mathbf{C}^* \cup \{C\}$ is an *extended 2-complete circ basis*. Since $\mathbf{C}^*$ is 2-complete it follows that each element of E belongs to *exactly* two members of $\mathbf{C}^* \cup \{C\}$ and hence the members of $\mathbf{C}^* \cup \{C\}$ can be interpreted as vertices of a graph. A binary graphoid with a 2-complete circ basis is called a *cographic binary graphoid*. Notice that there are binary graphoids with no 2-complete circ bases. For example this is the case with the binary graphoid $\mathcal{G}_2$ defined earlier. For this graphoid the subcollection $\mathbf{C}_I^* = \{\{a, b\}, \{a, c, d, e\}, \{d, e, f, g\}, \{b, e, f, h\}\}$ of the circ space $\mathbf{C}$ satisfies iii)$'$ but does not satisfy iv)$'$ since the element e belongs to three members of $\mathbf{C}_I^*$. Observe that the ring sum of all members of $\mathbf{C}_I^*$ is the subset $C_I = \{c, e, g, h\}$. If we interpret the members of $\mathbf{C}_I^* \cup C_I = \{\{a, b\}, \{a, c, d, e\}, \{d, e, f, g\}, \{b, e, f, h\}; \{c, e, g, h\}\}$ as vertices, the corresponding representation is not a graph but a hypergraph shown in Figure 2.8.

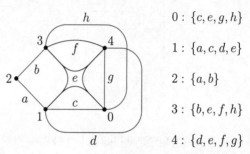

$0 : \{c, e, g, h\}$

$1 : \{a, c, d, e\}$

$2 : \{a, b\}$

$3 : \{b, e, f, h\}$

$4 : \{d, e, f, g\}$

Figure 2.8. A hypergraph corresponding to a circ base of the noncographic binary graphoid $\mathcal{G}_2$.

In this hypergraph every edge, except the edge e, is incident to a pair of distinct vertices; the edge e is incident to four vertices. Thus the binary graphoid $\mathcal{G}_2$ is both nongraphic and noncographic. Clearly, if a hypergraph corresponds to a circ basis of a binary graphoid then it also corresponds to a cut base of its dual and vice versa.

The binary graphoid induced by the circ and cut spaces of a graph is obviously graphic. However, if a binary graphoid is graphic it is not necessarily cographic. For example, see the binary graphoid induced by the circ space of the well known graph $K_{3,3}$ shown in Figure 2.9.

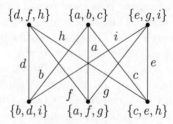

Figure 2.9. Graph $K_{3,3}$ illustrating a representation of a binary graphoid which is graphic but not cographic.

The fact that this binary graphoid is not cographic is intimately related to the following statements: *A binary graphoid is both graphic and cographic iff it is the graphoid associated with a planar graph.* It is easy to check that $K_{3,3}$ graph is not planar. On the other hand, the binary graphoid $\mathcal{G}_1$ defined earlier is both graphic and cographic since it is induced by the circ space of the graph of Figure 2.1 which is obviously planar.

As a conclusion, every graph satisfies (i), (ii) and (iii) and therefore can be seen as an interpretation of a binary graphoid in a particular 'coordinate system', defined by a 2-complete cut base. From this prospect, a binary graphoid is an essential, coordinate free, geometrical notion of which any associated graph is just a particular view of the same generality.

As we mentioned earlier, binary graphoids could be embedded into the larger class of general graphoids. Further motivation for introducing general graphoids (beside Corollaries 1.13 and 1.18 of Chapter 1 and the Orthogonality Theorem) can also be found in other statements about graphs. For example Corollary 2.13 and Remark 2.24.

Recall that a graphoid can be seen as a pair of dual matroids $\mathcal{M} = (E, \mathbf{C}_o)$ and $\mathcal{M}^* = (E, \mathbf{S}_o)$. We say that a subset I of E is an *independent set* of a matroid $\mathcal{M} = (E, \mathbf{C}_o)$ if I contains no members of $\mathbf{C}_o$. Otherwise, a subset is dependent. By $\mathbf{I}$ we denote the whole collection of independent sets of the matroid, that is, the collection $\mathbf{I} = \{I | C \not\subseteq I, \text{ for all } C \in \mathbf{C}_o\}$. Within matroid theory, it can be shown that the members of $\mathbf{I}$ satisfy the following conditions:

(I1) $\emptyset \in \mathbf{I}$.

(I2) If $X \in \mathbf{I}$ and $Y \subseteq X$ then $Y \in \mathbf{I}$.

(I3) If $X, Y \in \mathbf{I}$ and $|X| \leq |Y|$, then there exists $Z \in \mathbf{I}$ such that $X \subset Z \subseteq X \cup Y$.

The pair $(E, \mathbf{I})$ can be also used as another characterization of a matroid: if $\mathbf{I}$ is a collection of subsets of a finite set E, the members of which satisfy conditions (I1), (I2) and (I3), then the minimal subsets of E that do not belong to $\mathbf{I}$ satisfy conditions (C1) and (C2) of Corollary 1.18, Chapter 1. That is, a pair $(E, \mathbf{C}_o)$, uniquely determines the pair $(E, \mathbf{I})$, and vice versa. Thus, minimal dependent subsets are circuits and conversely, circuit-less subsets are independent sets.

A maximal independent set of a matroid is called a *base*. By B we denote a base of a matroid and by $\mathbf{B}$ the set of all bases of a matroid. Then, from matroid theory, the members of $\mathbf{B}$ satisfy the following conditions (which are the same as those for trees of graphs (see Remark 2.24)):

(B1) No member of $\mathbf{B}$ contains another member as a proper subset.

(B2) If $B_1, B_2 \in \mathbf{B}$, and $x_1 \in B_1 \setminus B_2$, then there exists $x_2 \in B_2 \setminus B_1$ such that $(B_1 \setminus \{x_1\}) \cup \{x_2\}) \in \mathbf{B}$.

Every independent set is simply a subset of a base and every base is a maximal independent set. The pair $(E, \mathbf{B})$ can be also used as a characterization of a matroid. This follows from the fact that if $\mathbf{B}$ is a collection of subsets of a finite set E that satisfy conditions (B1) and (B2), then the collection $\mathbf{I} = \{I | I \subseteq B, \text{ for some } B \in \mathbf{B}\}$ of all subsets of the members of $\mathbf{B}$ satisfies conditions (I1), (I2) and (I3).

For example, for the nongraphic matroid $\mathcal{M} = (E, \mathbf{C}_o)$ defined earlier as a circuit matroid on the set

$$E = \{a, b, c, d, e, f, g, h\}$$

where

$$
\begin{aligned}
\mathbf{C}_o \;=\; & \{\{a, b\}, \{b, c, d, e\}, \{a, c, d, e\}, \{b, c, f, g\}, \{a, c, f, g\}, \{d, e, f, g\}, \\
& \{b, d, g, h\}, \{a, d, g, h\}, \{c, e, g, h\}, \{c, d, f, h\}, \\
& \{b, e, f, h\}, \{a, e, f, h\}\},
\end{aligned}
$$

all its bases are contained in the following family of subsets

$$
\begin{aligned}
\mathbf{B} \;=\; & \{\{b, c, e, f\}, \{a, c, e, f\}, \{c, d, e, f\}, \{b, d, e, f\}, \{a, d, e, f\}, \\
& \{b, c, d, f\}, \{a, c, d, f\}, \{c, d, f, g\}, \{b, d, f, g\}, \{a, d, f, g\}, \\
& \{c, e, f, g\}, \{b, e, f, g\}, \{a, e, f, g\}, \{b, c, e, g\}, \{a, c, e, g\}, \\
& \{c, d, e, g\}, \{b, d, e, g\}, \{a, d, e, g\}, \{b, c, d, g\}, \{a, c, d, g\}, \\
& \{b, c, g, h\}, \{a, c, g, h\}, \{c, d, g, h\}, \{d, e, g, h\}, \\
& \{b, e, g, h\}, \{a, e, g, h\}, \{e, f, g, h\}, \{d, f, g, h\}, \\
& \{c, f, g, h\}, \{b, f, g, h\}, \{a, f, g, h\}, \{b, c, f, h\}, \{a, c, f, h\}, \\
& \{b, d, f, h\, \{a, d, f, h\}, \{d, e, f, h\}, \\
& \{c, e, f, h\}, \{b, c, e, h\}, \{a, c, e, h\}, \{c, d, e, h\}, \\
& \{b, d, e, h\}, \{a, d, e, h\}, \{b, c, d, h\}. \{a, c, d, h\}\}.
\end{aligned}
$$

If B is a base of a matroid $\mathcal{M} = (E, \mathbf{C}_o)$, then $B^* = E \setminus B$ is a base of its dual $\mathcal{M}^*$. The base of $\mathcal{M}^*$ is called a *cobase*. Therefore $\mathcal{M} = (E, \mathbf{C}_o)$ uniquely defines its dual $\mathcal{M}^* = (E, \mathbf{S}_o)$ and hence it also uniquely defines the graphoid $\mathcal{G} = (E, \mathbf{C}_o, \mathbf{S}_o)$.

The following is an addendum to the list of equivalent conditions: (0), (1) and (2), concerning circuit binary matroids, which was presented in the first section of this chapter:

(0) $\mathcal{M}$ *is a circuit binary matroid.*

(3) *For every base B and every circuit C of $\mathcal{M}$ such that $C \backslash B$ is nonempty, $C = \bigoplus_{e \in C \backslash t} C(e, b)$, where $C(e, b)$ is the fundamental circuit that $e \in C \backslash B$ forms with edges in B only (see also Lemma 2.28).*

(4) *For any two bases B' and B'' of $\mathcal{M}$ and any $y \in B''$, there are an odd number of members $x \in B'$ such that $(B' \backslash \{x\}) \cup \{y\}$ and $(B'' \backslash \{y\}) \cup \{x\}$ are bases of $\mathcal{M}$ (see also Corollary 2.25).*

Statement (3) is the same as the statement of Lemma 2.28. It is true for both graphs and binary matroids but not for general matroids. On the other hand, the statement of Corollary 2.29 (a corollary to Lemma 2.28) is true for general matroids.

2.6 Orientable binary graphoids and Kirchhoff's laws

As we pointed out in Proposition 1.40, given a graph G with edge set E of cardinality b, and given any oriented circuit C and any oriented cutset S of G, it is always true that *for every orientation of edges of G* and *every orientaion of C and S*, the scalar product $p^T q$ is equal to 0. Here p denotes the row incidence matrix of C against elements of E and q denotes the row incidence matrix of S against elements of E. Both p and q have entries that are 0, -1 and 1. The columns in both matrices are taken in the same order. Let $\mathbf{C}_o$ be the set of all circuits and let $\mathbf{S}_o$ be the set of all cutsets of G. Let P_o and Q_o be the associated incidence matrices. These matrices have 0, -1 and 1 entries. Then as an immediate consequence it follows that *for an arbitrary orientation of edges of G* and *an arbitrary orientation of circuits and cutsets*, the scalar product of any row of P_o and any row of Q_o is zero, providing that the columns in both matrices are taken in the same order. In the case of binary matroids this property, which plays a central role in the formulation of Kirchhoff's laws, generally does not hold. Instead, a similar property holds for the unoriented case involving matrices with 0 and 1 entries and *mod 2* arithmetic.

The following proposition for binary matroids (which is directly related to the Orthogonality Theorem) holds.

Proposition 2.39 *Given a binary graphoid $\mathcal{G} = (E, \mathbf{C}, \mathbf{S})$ with edge set E of cardinality b, let C be a circuit and let S be a cutset of $\mathcal{G}$. Denote by p the row matrix that describes the incidence of the circuit C against elements of E such that:*

$$p_k = \begin{cases} 1 & \textit{if } C \textit{ contains the edge } k, \\ 0 & \textit{if } C \textit{ does not contain the edge } k. \end{cases}$$

Denote by q the row matrix that describes the incidence of the cutset S against elements of E such that:

$$q_k = \begin{cases} 1 & \text{if } S \text{ contains the edge } k, \\ 0 & \text{if } S \text{ does not contain the edge } k. \end{cases}$$

Suppose that columns in both matrices are taken in the same order. Then the following scalar product is zero:

$$p^{\mathrm{T}}q = \sum_{k=1}^{b} p_k q_k = 0,$$

by means of mod 2 arithmetic.

Proof According to the Orthogonality Theorem the intersection $C \cap S$ is of even cardinality. If $C \cap S = \emptyset$ then for each $k \in \{1, \ldots, b\}$, at least one of p_k and q_k is equal to zero. $\sum_{k=1}^{b} p_k q_k = 0$. Suppose now that $C \cap S$ is not empty. Then for an even nonzero number of edges k, both p_k and q_k are nonzero and hence the sum contains an even nonzero number of equal entries (all equal to 1). Accordingly, $\sum_{k=1}^{b} p_k q_k = 0$, by means of mod 2 arithmetic. $\qquad\square$

For example, for the circuit $\{d, e, h, i\}$ and the cutset $\{c, e, h\}$ of the graph of Figure 2.9 the associated circuit row incidence matrix p and cutset incidence matrix q respectively are:

$$p = \begin{array}{c} \\ \{d,e,h,i\} \end{array} \begin{array}{ccccccccc} [a] & [b] & [c] & [d] & [e] & [f] & [g] & [h] & [i] \\ 0 & 0 & 0 & 1 & 1 & 0 & 0 & 1 & 1 \end{array},$$

$$q = \begin{array}{c} \\ \{c,e,h\} \end{array} \begin{array}{ccccccccc} [a] & [b] & [c] & [d] & [e] & [f] & [g] & [h] & [i] \\ 0 & 0 & 1 & 0 & 1 & 0 & 0 & 1 & 0 \end{array}.$$

Clearly $p^{\mathrm{T}}q = 1 + 1 = 0 (mod\ 2)$.

The next corollary is an immediate consequence of Proposition 2.39:

Corollary 2.40 *Let $\mathcal{G} = (E, \mathbf{C}, \mathbf{S})$ be a binary graphoid with edge set E and let $\mathbf{C}_o$ be the set of all its circuits and let $\mathbf{S}_o$ be the set of all its cutsets. Let P_o be a matrix describing the incidence of members of $\mathbf{C}_o$ against the members of E whose entries for $i = 1, \ldots, |\mathbf{C}_o|$ and $k = 1, \ldots, |E|$ are:*

$$p_{ik} = \begin{cases} 1 & \text{if circuit } i \text{ contains edge } k, \\ 0 & \text{if circuit } i \text{ does not contain edge } k. \end{cases}$$

and let Q_o be a matrix describing the incidence of members of $\mathbf{S}_o$ against the members of E whose entries for all $j = 1, \ldots, |\mathbf{S}_o|$ and $k = 1, \ldots, |E|$ are:

$$q_{jk} = \begin{cases} 1 & \text{if cutset } i \text{ contains edge } k, \\ 0 & \text{if cutset } i \text{ does not contain edge } k. \end{cases}$$

Suppose that the columns in both matrices are taken in the same order. Then, the scalar product of any row of P_o and any row of Q_o is zero, by means of mod 2 arithmetic.

We call a binary graphoid an $\mathcal{G} = (E, \mathbf{C}, \mathbf{S})$ *orientable binary graphoid* (or simply a *regular graphoid*) if it is possible to assign negative signs to some of the nonzero entries of P_o and Q_o in such way that the scalar product of any row of P_o and any row of Q_o is 0, where now we use conventional integer arithmetic unlike *mod* 2 arithmetic used previously.

The following properties concerning orientability of binary graphoids hold:

(i) *A graphoid is orientable iff its dual is orientable.*

(ii) *Any graphic and any cographic graphoid is orientable.*

(o) *Not every binary graphoid is orientable.*

Suppose $\mathcal{G} = (E, \mathbf{C}, \mathbf{S})$ is an orientable binary graphoid and let $\mathbf{C}^*$ be one of its circ bases and $\mathbf{S}^*$ be one of its cut bases. Consider the hypergraphs that corresponds to these bases. In our representation of hypergraphs, the arcs joining hyper edges to vertices are called *terminals*. Every terminal can be directed so as to enter or to leave a vertex. Now suppose that all edges of the hypergraphs are oriented in the following way: every terminal of each hyper edge is oriented so that total number of terminal orientations that enter vertices is equal to the total number of those that leave vertices (notice that the number of terminals of any edge in the hypergraph corresponding to a binary graphoid is even). For two terminal edges it is sufficient to orient only one of its terminal. For the hypergraph whose vertices correspond to the extention of the circ basis $\mathbf{C}^*$, denote by P the matrix that describes the incidence of the oriented vertices of this hypergraph (these correspond to members of $\mathbf{C}^*$) against oriented edges (these correspond to members of E). Similarly, for the hypergraph whose vertices correspond to the extension of the basis $\mathbf{S}^*$, denote by Q be the matrix that describes the incidence of the oriented vertices (these correspond to members of $\mathbf{S}^*$) against oriented edges (these correspond to members of E). Both are matrices with entries $1, -1$ or 0.

Denote by u the $(b \times 1)$ column matrix whose components are voltages and denote by i the $(b \times 1)$ column matrix whose components are currents. Assume that rows in both are taken in the same order and that this order coincides with the order of the columns of matrices P and Q. Then the set of equations based on Kirchhoff's laws written for a collection of circuits $\mathbf{C}^*$ and a collection of cutsets $\mathbf{S}^*$ can be described by the following matrix equations:

$$Pu = 0 \tag{a}$$

$$Qi = 0. \tag{b}$$

From the point of view of network analysis, despite normal practice, the necessity of using a digraph to describe the topology of a network is just a prejudice. Denote by E the set of all network ports. We associate a current and a voltage with each oriented port. Kirchhoff's laws (see section 1.9), as constitutional laws for topological considerations, do not really need a digraph. What are really essential are two families of port subsets: the family of all oriented circs $\mathbf{C}$ and the family of all oriented cuts $\mathbf{S}$. In other words, Kirchhoff's laws only require an orientable graphoid $(E, \mathbf{C}, \mathbf{S})$. As we mentioned before, several graphs may have the same graphoid. A graph is an over-detailed description of the network topology. Suppose we consider a class of 2-isomorphic graphs each describing a network topology. Since the set of all circs $\mathbf{C}$ and the set of all cuts $\mathbf{S}$ are the same for all these graphs (they all have the same triple $(E, \mathbf{C}, \mathbf{S})$) it follows that they all describe the same Kirchhoff's space (the set of all solutions of the system of equations based on Kirchhoff's voltage and current laws) and therefore describe the same network topology. For example, the two graphs of Figure 2.1 are distinct and yet describe the same network topology.

As an example of an orientable binary graphoid consider the graphoid associated with the well known $K_{3,3}$ graph (which is graphic but not cographic) oriented as shown in Figure 2.11. Its nonoriented counterpart was shown earlier in Figure 2.9.

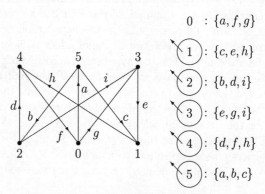

$$0 \; : \{a, f, g\}$$
$$\boxed{1} : \{c, e, h\}$$
$$\boxed{2} : \{b, d, i\}$$
$$\boxed{3} : \{e, g, i\}$$
$$\boxed{4} : \{d, f, h\}$$
$$\boxed{5} : \{a, b, c\}$$

Figure 2.10. Graph $K_{3,3}$ oriented.

It can be seen by inspection that $\mathbf{S}^* = \{\{c, e, h\}, \{b, d, i\}, \{e, g, i\}, \{d, f, h\}, \{a, b, c\}\}$ is a cut basis of this graph and that $\mathbf{S}^* \cup \{a, f, g\}$, is the associated extended basis. Notice that $|E| = 9$ and $|\mathbf{S}^*| = n = 5$. The corresponding matrix Q which describes incidence of independent cutsets of this graph (members of $\mathbf{S}^*$) against its edges is:

$$Q = \begin{array}{c|ccccccccc} & [a] & [b] & [c] & [d] & [e] & [f] & [g] & [h] & [i] \\ \{c,e,h\} & 0 & 0 & -1 & 0 & -1 & 0 & 0 & 1 & 0 \\ \{b,d,i\} & 0 & -1 & 0 & 1 & 0 & 0 & 0 & 0 & 1 \\ \{e,g,i\} & 0 & 0 & 0 & 0 & 1 & 0 & -1 & 0 & -1 \\ \{d,f,h\} & 0 & 0 & 0 & -1 & 0 & 1 & 0 & -1 & 0 \\ \{a,b,c\} & -1 & 1 & 1 & 0 & 0 & 0 & 0 & 0 & 0 \end{array}.$$

Then Kirchhoff's current laws are described by the following equations:

$$
\begin{aligned}
-i_c - i_e + i_c + i_h &= 0 \\
-i_b + i_d + i_i &= 0 \\
i_e - i_g - i_i &= 0 \\
-i_d + i_f - i_h &= 0 \\
-i_a + i_b + i_c &= 0.
\end{aligned}
$$

It can be also seen by inspection from Figure 2.11 that $\mathbf{C}^* = \{\{e,f,g,h\},$ $\{d,e,h,i\}, \{b,c,d,h\}, \{a,c,f,h\}\}$ is a circ basis of this graph and that $\mathbf{C}^* \cup \{a,b,g,i\}$, is the associated extended basis. Since $K_{3,3}$ is not planar it follows that there is no circ basis $\mathbf{C}^*$ which is 2-complete. Hence every representation of the dual of $K_{3,3}$ graph is a hypergraph. An oriented hypergraph associated with the circ basis is shown in Figure 2.10.

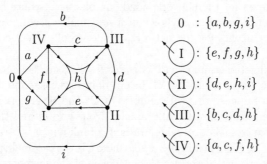

Figure 2.11. A hypergraph which corresponds to a circ basis of $K_{3,3}$ graph.

Notice that $|E| = 9$ and $|\mathbf{C}^*| = m = 4$. The corresponding matrix P which describes the incidence of independent circuits of this graph (members of $\mathbf{C}^*$) against its edges is:

$$P = \begin{array}{c|ccccccccc} & [a] & [b] & [c] & [d] & [e] & [f] & [g] & [h] & [i] \\ \{e,f,g,h\} & 0 & 0 & 0 & 0 & -1 & -1 & -1 & -1 & 0 \\ \{d,e,h,i\} & 0 & 0 & 0 & -1 & 1 & 0 & 0 & 1 & 1 \\ \{b,c,d,h\} & 0 & 1 & -1 & 1 & 0 & 0 & 0 & -1 & 0 \\ \{a,c,f,h\} & 1 & 0 & 1 & 0 & 0 & 1 & 0 & 1 & 0 \end{array}.$$

Then Kirchhoff's voltage laws are described by the following equations:

$$-u_e - u_f - u_g - u_h = 0$$
$$-u_d + u_e + u_h + u_i = 0$$
$$u_b - u_c + u_d - u_h = 0$$
$$u_a + u_c + u_f + u_h = 0.$$

Obviously, matrices P and Q are orthogonal.

2.7   Mesh and nodal analysis

Independent structures of oriented graphs play an important role in network analysis. Thus, every maximal independent collection of oriented circuits (cutsets) serves as a basis for the formulation of maximal linearly independent system of equations based on Kirchhoff's voltage (current) laws. On the other hand every maximal collection of oriented circuit-less (cutset-less) edges serves as a basis for determining a maximal collection of topologically independent voltages (currents).

Nodal and mesh analyses are classical techniques for the analysis of electrical networks in terms of topologically independent set of variables. The former is based on a choice of a maximal collection of topologically independent voltages for the networks made up of voltage-controlled ports only, and the latter is based on a choice of a maximal collection of topologically independent currents for the networks made up of current-controlled ports only.

Given a directed graph G with edge set E, let $\mathbf{C}$ and $\mathbf{S}$ be its collections of circs and cuts, respectively. Let $\mathbf{C}^*$ be one of its circ bases and let $\mathbf{S}^*$ be one of its cut bases. Suppose that all members of $\mathbf{C}^*$ and $\mathbf{S}^*$ are oriented. Suppose also that the members of $\mathbf{C}^*$ are labelled from 1 to m and the members of $\mathbf{S}^*$ are labelled from 1 to n, where m is a corank and n is a rank of G. The members of E are labelled from 1 to b, where b is the cardinality of E. Denote by P the matrix that describes the incidence of the members of $\mathbf{C}^*$ against the members of E of the directed graph G whose entries for all $i = 1, \ldots, m$ and $k = 1, \ldots, b$ are:

$$p_{ik} = \begin{cases} 1 & \text{if the circuit } i \text{ contains edge } k \text{ and their orientations coincide,} \\ -1 & \text{if the circuit } i \text{ contains edge } k \text{ and their orientations are} \\ & \hspace{9cm} \text{opposite,} \\ 0 & \text{if the circuit } i \text{ does not contain edge } k. \end{cases}$$

Denote by Q the matrix that describes the incidence of the members of $\mathbf{S}^*$ against the members of E of the directed graph G whose entries for all $j =$

$1, \ldots, n$ and $k = 1, \ldots, b$ are

$$q_{jk} = \begin{cases} 1 & \text{if the cutset } i \text{ contains edge } k \text{ and their orientations coincide,} \\ -1 & \text{if the cutset } i \text{ contains edge } k \text{ and their orientations are} \\ & \hspace{7cm} \text{opposite,} \\ 0 & \text{if the cutset } i \text{ does not contain edge } k. \end{cases}$$

Thus both matrices P and Q are matrices with 1, -1 or 0 entries which we call a *reduced circuit incidence matrix* and a *reduced cutset incidence matrix*, respectively. In the matrix P the oriented circuits are associated with the rows, and oriented edges with the columns. In the matrix Q the oriented cutsets are associated with the rows, and oriented edges with the columns. Clearly, according to Corollary 1.41, Chapter 1, P and Q are orthogonal, that is, $PQ^{\mathrm{T}} = 0_{m,n}$ and $QP^{\mathrm{T}} = 0_{n,m}$, where $0_{m,n}$ and $0_{n,m}$ are $(m \times n)$ and $(n \times m)$ matrices respectively with zero entries.

Given an oriented graph with b edges, denote by u the $(b \times 1)$ vector whose components are port voltages and denote by i the $(b \times 1)$ vector whose components are port currents. Assume that rows in both column matrices are taken in the same order and that this coincides with the order of columns of matrices P and Q. Then the maximal linearly independent equations based on Kirchhoff's laws are described by the following equations:

$$Pu = 0_{m,1} \tag{1}$$

$$Qi = 0_{n,1}. \tag{2}$$

For example, for the oriented graph of Figure 1.28 the associated circuit incidence matrix P and cutset incidence matrix Q respectively are:

$$P = \begin{array}{c} \\ \{a,b,d,e\} \\ \{c,d,e\} \\ \{e,f\} \end{array} \begin{array}{cccccc} [a] & [b] & [c] & [d] & [e] & [f] \\ 1 & 1 & 0 & 1 & -1 & 0 \\ 0 & 0 & 1 & -1 & 1 & 0 \\ 0 & 0 & 0 & 0 & -1 & 1 \end{array},$$

$$Q = \begin{array}{c} \\ \{a,b\} \\ \{b,c,e,f\} \\ \{d,e,f\} \end{array} \begin{array}{cccccc} [a] & [b] & [c] & [d] & [e] & [f] \\ 1 & -1 & 0 & 0 & 0 & 0 \\ 0 & 1 & -1 & 0 & 1 & 1 \\ 0 & 0 & 0 & -1 & -1 & -1 \end{array}.$$

Obviously, P and Q are orthogonal.

Let v be an $(n \times 1)$ vector whose components are independent port voltages (*generalized voltage potentials*) associated with the members of $\mathbf{S}^*$. Then Kirchhoff's voltage law (KVL) can be described by the following equation:

$$u = Q^{\mathrm{T}} v. \tag{1'}$$

This is KVL in so-called explicit form. A vector u satisfies the relation (1) iff there is a vector v such that $(1')$ holds. In fact, $(1')$ describes the set of all solutions of equation (1), since v is a vector with arbitrary entries.

Denote by j an $(m \times 1)$ vector whose components are independent port currents (*generalized current potentials*) associated with the members of $\mathbf{C}^*$. Then Kirchhoff's current law (KCL) can be described by the following equation:

$$i = P^{\mathrm{T}} j. \tag{2'}$$

This is KCL in so-called explicit form. A vector i satisfies the relation (2) iff there is a vector j such that $(2')$ holds. In fact, $(2')$ describes the set of all solutions of equation (2), since j is a vector with arbitrary entries.

Proposition 2.41 [The Theorem of Tellegen] *If u satisfies relation (1) and u satisfies relation (2) then $u^{\mathrm{T}} i = 0$.*

Proof From $(1')$ and (2) it follows that $u^{\mathrm{T}} i = (Q^{\mathrm{T}} v)^{\mathrm{T}} i = v^{\mathrm{T}}(Q i) = 0$. $\square$

Any two of the following three relations $P u = 0_{m,1}$, $Q i = 0_{n,1}$ and $u^{\mathrm{T}} i = 0$, imply the third. This can be proved using similar arguments to those for the proof of Proposition 2.41.

Suppose now that all multiports in the network are voltage-controlled:

$$i = \hat{i}(u). \tag{3}$$

If we substitute (2) into (3) then the port currents are completely controlled by independent port voltages only:

$$i = \hat{i}(Q^{\mathrm{T}} v). \tag{3'}$$

Substituting $(3')$ into (2), we finally obtain $|\mathbf{S}^*| = n$ equations in terms of $|\mathbf{S}^*| = n$ independent port voltages:

$$Q \hat{i}(Q^{\mathrm{T}} v) = 0$$

which we call *nodal equations* of resistive networks. *Thus, the number of nodal equations of a network is equal to the rank of the graph which describes the topology of the network.*

Suppose now that all multiports in the network are current-controlled:

$$u = \hat{u}(i). \tag{4}$$

If we substitute $(2')$ into (4) then the port voltages are completely controlled by independent port currents only:

$$u = \hat{u}(P^{\mathrm{T}} j). \tag{4'}$$

Substituting (4') into (1), we finally obtain $|\mathbf{C}^*| = m$ equations in terms of $|\mathbf{C}^*| = m$ independent port currents:

$$P\hat{u}(P^T j) = 0$$

which we call *mesh equations* of resistive networks. *Thus, the number of mesh equations of a network is equal to the corank of the graph which describes the topology of the network.*

Nodal and mesh equations are dual network equations by means of the following correspondences: *voltages* ↔ *currents, circuits* ↔ *cutsets, bases* ↔ *cobases, fundamental circuits* ↔ *fundamental cutsets*. When some ports are v-controlled and some are i-controlled, neither of the nodal and mesh equations can be used for network analysis. In this case a system of *hybrid network equations* expressed in terms of a mixture of voltage and current variables is needed (see the last section of Chapter 3).

2.8   Bibliographic notes

The comparative expositions of graphs and matroids in the paper of Harary and Welsh [28] and the books of Recski [61], Swammy and Thulasiraman [65], Wilson [77], and Tutte [69], provide deep insight. Whitney first used the term matroid in his famous paper on the abstract properties of linear dependence [75]. Minty [46] established a unified approach to graphoids using the Painting Theorem. The statement of Corollary 2.13 is the well-known Tutte Lemma [68]. For more on independence properties and matroid axioms see Welsh [73], Tutte [69], Swamy and Thulasiraman [65], Bryant and Perfect [6], Fournier [18] and Recski [61]. Some recent results on fuzzy matroids and fuzzy independence set systems can be found in the papers of Goetschel and Voxman [22], [23] and Novak [51]. Early papers on graphs in the context of vector spaces are by Gould [24], Chen, W-K. [9] and Williams and Maxwell [76]. For some aspect of representing graphs via incidence matrices see also [42]. For the use of orientable binary matroids in the context of circuit theory see the book of Sillamaa [64] (available only in Russian). For the connection between matroids and network analysis see also Weinberg [72] and a number of papers referenced therein. The matroidal version of Corollary 2.29 is presented in [49]. Corollaries 2.13, 2.18, 2.20 and 2.25 are graph versions of some well known matroidal statements (see Welsh [73], Tutte [69], and Swammy and Thulasiraman [65]). The catalogue of all non-isomorphic matroids on at most 8 elements is compiled in [2] by Acketa. The nonregular binary graphoid $\mathcal{G}_2$ which is both nongraphic and noncographic used in this chapter is taken from this catalogue. The matroidal version of the statement of Corollary 2.29 was proved by Novak [49].

3

Basoids

In this chapter maximal edge subsets that are both circuit-less and cutset-less (which we call basoids) and the related concepts of principal minor and principal partition of a graph are considered. The fact that basoids may have different cardinalities provides a rich structure which is described through several propositions. Transitions from one basoid to another (which provides a basis for augmenting basoids in turn) and the concept of a minor of a graph with respect to a dyad (a maximum cardinality basoid) are also investigated in detail. It is shown that there exists a unique minimal minor with respect to every dyad of a graph G. This edge subset, called the principal minor and its dual called the principal cominor, define a partition of the edge set of G called the principal partition. The hybrid rank of a graph is defined to be the cardinality of a dyad of the graph. This is a natural extension of the definitions of rank and corank of a graph, defined as the cardinalities of a maximum circuit-less subset and a maximum cutset-less subset of the graph, respectively. In the last section of this chapter an application to hybrid topological analysis of networks is considered. An algorithm for finding a maximum cardinality topologically complete set of network variables is also described.

The material of this chapter is general in the sense that it can be easily extended from graphs to matroids. To ensure this generality, the Painting Theorem, the matroidal version of the Orthogonality Theorem as well as the Circuit and the Cutset axioms are widely used to prove propositions.

3.1 Preliminaries

Familiarity with the basic concepts of graphs such as circuit and cutset are presumed in this chapter. A maximal circuit-less subset of a graph G is called a *tree* of G while a maximal cutset-less subset of edges is called a *cotree*. These terms (circuit, cutset, tree and cotree) will be used here to mean a subset of the edges of a graph. Let F be a subset of E. Then the rank of F, denoted by rank (F), is the cardinality of a maximum circuit-less subset of F and the corank of F, denoted by corank (F), is the cardinality of a maximum cutset-less subset of F. The complement of F is the set difference $E \setminus F$, denoted by F^*. By $|F|$ we denote the number of elements in (that is, the cardinality of) the subset F. The *span* of an edge subset F of a graph G with respect to circuits (respectively, cutsets), denoted by $sp_c(F)$ ($sp_s(F)$), is defined to be the union of F and all the circuits (cutsets) that individual edges in G form with edges in F only. Besides the usual set operations ($\cup$, $\setminus$, and $\cap$) we also use the set operation $\oplus$, called the ring sum. A collection of subsets of a finite set, closed under inclusion, is called an *independence set system*. A member of an independence set system is called an *independent set*. An independent set is maximal if no superset of it is also a member of the system. By presuming additional conditions on an independence set system we can obtain different subset system structures. An important example is the matroid structure (see Section 6 of Chapter 2). Two special matroids can be associated with every graph: the graphic matroid and the cographic matroid. The independent sets of the graphic matroid are circuit-less subsets of the graph while the independent sets of the cographic matroid are cutset-less subsets of the graph.

In proving results we occasionally refer to the following theorems which are well-known in the graph-theoretic and matroid literature.

The Orthogonality Theorem [Proposition 1.20] *Let C be a circuit and S be a cutset of a graph. Then $|C \cap S| \neq 1$.*

The Painting Theorem [Proposition 1.27] *Given a graph G with the edge set E, let e be a member of E that forms neither a self-circuit nor a self-cutset and let $(E_1, \{e\}, E_2)$ be a partition of E. Then either e forms a circuit with members of E_1 only or a cutset with members of E_2 only, but not both.*

The Cutset axiom [Corollary 1.13] *Let S_1 and S_2 be two cutsets of a graph with nonempty intersection and let $e \in S_1 \cap S_2$. Let also $x \in S_1 \setminus S_2$. Then, there is a cutset S of the graph such that $x \in S \subseteq (S_1 \cup S_2) \setminus \{e\}$.*

The Circuit axiom [Corollary 1.18] *Let C_1 and C_2 be two circuits of a graph with nonempty intersection and let $e \in C_1 \cap C_2$. Let also $x \in C_1 \setminus C_2$. Then, there is a circuit C of the graph such that $x \in C \subseteq (C_1 \cup C_2) \setminus \{e\}$.*

3.2   Basoids of graphs

A subset of edges of a graph G is said to be a *double independent subset* if it contains no circuits and no cutsets of G. The collection of all double independent subsets of a graph is obviously closed under inclusion and therefore constitutes an independence set system. Within the context of matroid theory double independent edge subsets can be regarded as subsets which are independent in two matroids: the graphic matroid and its dual, the co-graphic matroid. This is why simultaneous circuit-less and cutset-less edge subsets are called double independent subsets of a graph. A double inde-pendent subset of edges which is maximal in the sense that no other double independent subset of G contains it as a proper subset, we call a *basoid*. The name basoid for a maximal double independent edge subset of a graph is used to distinguish it from a maximal independent subset of a matroid which is called a base of a matroid. In contrast to maximal circuit-less subsets (trees) or to maximal cutset-less subsets (cotrees) of a graph, maximal subsets that are both circuit-less and cutset-less (basoids) have not necessarily the same cardinality. We adopt the name *dyad* for maximum double independent edge subsets, that is for largest basoids.

In matroids, maximal independent subsets all have the same cardinality and hence, every maximal independent subset is at the same time a maximum independent subset. The following three examples show that basoids do not all have the same cardinality.

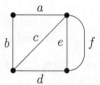

Figure 3.1. A simple graph with basoids of two different cardinalities.

Example 3.1 The basoids of the graph of Figure 3.1 can be placed in two classes according to their cardinality. The basoid $\{c, d\}$ has cardinality 2 whereas all the others $\{b, c, e\}$, $\{b, c, f\}$, $\{b, d, e\}$, $\{b, d, f\}$, $\{a, c, e\}$, $\{a, c, f\}$, $\{a, e, d\}$, $\{a, f, d\}$ have cardinality 3. This is one of the smallest examples with basoids of at least two different cardinalities.

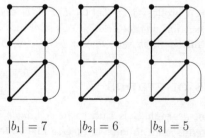

$|b_1| = 7$ $|b_2| = 6$ $|b_3| = 5$

Figure 3.2. An example of a graph with basoids of three different cardinalities.

Example 3.2 Figure 3.2 shows three copies of an augmented version of the graph of Figure 3.1. All basoids for this example have cardinality 7, 6 or 5. Representative basoids b_1, b_2 and b_3, one for each cardinality, are indicated in the figure. The graph of Figure 3.2 has 128 basoids with cardinality 7, 80 with cardinality 6 and 8 with cardinality 5.

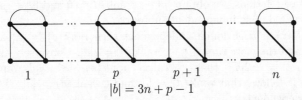

$$|b| = 3n + p - 1$$

Figure 3.3. An example for the construction of graphs with basoids having a range of cardinalities.

Example 3.3 Taking the graph of Example 3.1 as a basic unit, consider the class of graphs obtained by stringing together a number of these units. For example, Figure 3.2 shows the string of two units. Figure 3.3 shows the string of n units. Such a graph has a set of basoids which can be classified into $n+1$ subsets having cardinalities $3n - 1, 3n, \ldots, 4n - 1$.

We now present some properties of double independent subsets and basoids that are directly connected with the Painting Theorem. The first proposition provides necessary and sufficient conditions for a subset of edges to be a double independent subset.

Proposition 3.4 *Given an edge subset d of a graph G the following statements are equivalent:*

(i) *d is a double independent subset of G.*

(ii) *Every edge in d forms both a circuit and a cutset of G with edges in d^* only.*

(iii) $\mathrm{rank}\,(d^*) = \mathrm{rank}\,(E)$ *and* $\mathrm{corank}\,(d^*) = \mathrm{corank}\,(E)$.

(iv) $\mathrm{rank}\,(d) + \mathrm{corank}\,(d) = 2|d|$.

Proof (i) $\Leftrightarrow$ (ii) Let e belong to d. Since e forms neither a circuit nor a cutset with edges in d, according to the Painting Theorem applied to the triple $(d\backslash\{e\}, \{e\}, d^*)$, we conclude that e must form both a circuit and a cutset with edges in d^* only. Conversely, if every edge in d forms both a circuit and a cutset with edges in d^* only then, again according to the Painting Theorem, no edge in d forms a circuit or a cutset with edges in d only.

(i) $\Leftrightarrow$ (iii) As is well known (see Corollary 2.27 of Chapter 2), the relations $|F| - \operatorname{corank}(F) = \operatorname{rank}(E) - \operatorname{rank}(F^*)$ and $|F| - \operatorname{rank}(F) = \operatorname{corank}(E) - \operatorname{corank}(F^*)$ hold for every $F \subseteq E$. Let d be a double independent subset of G. Then, because $|d| = \operatorname{rank}(d) = \operatorname{corank}(d)$, setting $F = d$ in above relations, we obtain $0 = \operatorname{rank}(E) - \operatorname{rank}(d^*)$ and $0 = \operatorname{corank}(E) - \operatorname{corank}(d^*)$. Conversely, setting $F = d$, $\operatorname{rank}(d^*) = \operatorname{rank}(E)$ and $\operatorname{corank}(d^*) = \operatorname{corank}(E)$ in the above relations, we obtain $|d| - \operatorname{corank}(d) = 0$ and $|d| - \operatorname{rank}(d) = 0$, that is, d is a double independent subset of G.

(i) $\Leftrightarrow$ (iv) If d is double independent then $\operatorname{rank}(d) = |d|$ and $\operatorname{corank}(d) = |d|$ and therefore $\operatorname{rank}(d) + \operatorname{corank}(d) = 2|d|$. Conversely, if $\operatorname{rank}(d) + \operatorname{corank}(d) = 2|d|$ then because $\operatorname{rank}(d) \leq |d|$ and $\operatorname{corank}(d) \leq |d|$ it follows that $\operatorname{rank}(d) = |d|$ and $\operatorname{corank}(d) = |d|$. That is, d is double independent. $\square$

The next proposition serves as a basis for augmenting a double independent edge subset until a basoid is obtained.

Proposition 3.5 *Let d be a double independent edge subset of a graph G and let $e \in d^*$. Then, the following statements are equivalent:*

(i) *$d \cup \{e\}$ is double independent in G;*

(ii) *e belongs to both a circuit and a cutset of G made from edges in d^* only;*

(iii) *$\operatorname{rank}(d^* \setminus \{e\}) = \operatorname{rank}(d^*)$ and $\operatorname{corank}(d^* \setminus \{e\}) = \operatorname{corank}(d^*)$.*

Proof (i)$\Leftrightarrow$ (ii) According to the Painting Theorem applied to the triple $(d, \{e\}, d^* \setminus \{e\})$, an edge e belongs to both a circuit and a cutset made from edges in d^* iff e does not form either a circuit or a cutset with edges in d only. Because d is double independent, this means that $d \cup \{e\}$ is also double independent.

(i)$\Rightarrow$ (iii) Suppose both d and $d \cup \{e\}$ are double independent in G. Then, according to Proposition 3.4, $\operatorname{rank}(d^*) = \operatorname{rank}(E)$, $\operatorname{corank}(d^*) = \operatorname{corank}(E)$, $\operatorname{rank}((d \cup \{e\})^*) = \operatorname{rank}(E)$ and $\operatorname{corank}((d \cup \{e\})^*) = \operatorname{corank}(E)$. Thus, $\operatorname{rank}(d^*) = \operatorname{rank}((d \cup \{e\})^*) = \operatorname{rank}(d^* \setminus \{e\})$ and $\operatorname{corank}(d^*)\operatorname{corank}((d \cup \{e\})^*) = \operatorname{corank}(d^* \setminus \{e\})$.

(iii)$\Rightarrow$ (i) Suppose $\operatorname{rank}(d^* \setminus \{e\}) = \operatorname{rank}(d^*)$ and $\operatorname{corank}(d^* \setminus \{e\}) = \operatorname{corank}(d^*)$ for every edge $e \in d^*$. Because every subset of a double independent subset is also double independent, deleting edges of d in turn until the empty set is attained, we obtain a sequence $d = d_0, d_1, \ldots, d_n = \emptyset$ where $n = |d|$, $\operatorname{rank}((d_0)^*) = \operatorname{rank}((d_1)^*) = \ldots = \operatorname{rank}((d_n)^*)$. But $(d_0)^* = d^*$ and $(d_n)^* = (\emptyset)^* = E$, and therefore $\operatorname{rank}(d^*) = \operatorname{rank}(E)$ and $\operatorname{corank}(d^*) = \operatorname{corank}(E)$. $\square$

Notice that in proving the equivalence (i)⇔ (ii) of Proposition 3.4, the Painting Theorem was applied to the triple $(d \setminus \{e\}, \{e\}, d^*)$, whereas in proving the equivalence (i)⇔(ii) of Proposition 3.5 it was applied to the triple $(d, \{e\}, d^* \setminus \{e\})$.

Statements analogous to Propositions 3.4 and 3.5, for circuit-less (respectively, cutset-less) subsets, can be formulated as follows. *A subset g of the edges of a graph is circuit-less (cutset-less) iff each edge in g forms a cutset (circuit) with edges in g^* only.* This is an immediate consequence of the Painting Theorem applied to the triple $(g \setminus \{e\}, \{e\}, g^*)$ where e belongs to g. In an analogous manner, applying the Painting Theorem to the triple $(g, \{e\}, g^* \setminus \{e\})$, it can be proved that the following statements hold: *Let g be a circuit-less (cutset-less) edge subset of G and let e belong to g^*. Then $g \cup \{e\}$ is circuit-less (cutset-less) iff e belongs to a cutset (circuit) made from edges in g^* only.*

The following two algorithms are based on Proposition 3.5.

Algorithm 1 (To find a basoid)
Input: A graph G
begin
$f \leftarrow$ set of all edges of the graph G
$b \leftarrow \emptyset$
while f is nonempty **do**
begin
    Contract the edges in G that belong to b and, in the graph
    obtained, identify edges that belong to self-circuits. Denote
    this set by C_o. Remove the edges in G that belong to b and,
    in the resultant graph, identify edges that belong to self-cutsets.
    Denote this set by S_o
    $f \leftarrow (b \cup C_o \cup S_o)^*$
    Choose any edge $e \in f$
    $b \leftarrow b \cup \{e\}$
    end
output the basoid b
end of Algorithm 1

Algorithm 2 (To find a basoid)
Input: A graph G
begin
$G' \leftarrow G, G'' \leftarrow G, b \leftarrow \emptyset$
while corank (G') = corank (G) **and** rank (G'') = rank (G) **do**
    begin
    Choose an edge $e \in b^*$ that forms neither a self-circuit of G' nor

a self-cutset of G''. Contract e in G' and remove e from G'' and
denote the new graphs by G'_e and G''_e

$G' \leftarrow G'_e$, $G'' \leftarrow G''_e$, $b \leftarrow b \cup \{e\}$
   end
end of Algorithm 2

Let d be a double independent subset of a graph. Denote by $Q(d)$ the set
of edges in d^* that belong to circuits made of edges in d^* only, and denote by
$P(d)$ the set of edges in d^* that belong to cutsets made of edges in d^* only.
According to the Painting Theorem, an edge belongs to the set $Q(d)$ iff it
does not form a cutset with edges in d only. Dually, an edge belongs to the
set $P(d)$ iff it does not form a circuit with edges in d only. From the same
theorem, an edge belongs to both $Q(d)$ and $P(d)$, iff it forms neither a circuit
nor a cutset with edges in d only. Dually, an edge belongs to $d^* \setminus (Q(d) \cup P(d))$
iff it forms both a circuit and a cutset with edges in d only. Notice that in
contrast to d^*, every edge in d forms both a circuit and a cutset with edges
of d^* (Proposition 3.4).

The next proposition establishes necessary and sufficient conditions for a
double independent subset to be a basoid.

Proposition 3.6 *Given a double independent subset b of a graph G the fol-*
lowing statements are equivalent:

(i) *b is a basoid;*

(ii) *any circuit and any cutset of G made of edges in b^* only are disjoint;*

(iii) *any edge in b^* forms a circuit and/or a cutset of G with edges in b only.*

Proof (i)⟺ (ii) A double independent subset b is not a basoid if there is
an edge $e \in b^*$, such that $b \cup \{e\}$ is still double independent. According to
Proposition 3.5 this is true iff e belongs at the same time to a circuit and a
cutset made of edges in b^* only. Thus, (i) is not true iff (ii) is not true which
implies, (i) is true iff (ii) is true.

(ii)⟺ (iii) Statement (ii) is not true iff there is an edge $e \in b^*$ that forms
both a circuit and a cutset with edges in b^* only. By applying the Painting
Theorem to the triple $(b, \{e\}, b^* \setminus \{e\})$, we deduce that e forms neither a
cutset nor a circuit with edges in b only. Thus, (ii) is not true iff (iii) is not
true which implies, (ii) is true iff (iii) is true.   □

Remark 3.7 A statements analogous to Propositions 3.6, for maximal
circuit-less (cutset-less) subsets, can be formulated as follows: *Given a circuit-*
less (cutset-less) subset g of a graph G the following statements are equivalent:

(i) g is a maximal circuit-less (cutset-less) subset, of G;

(ii) there is no cutset (circuit) of G made of edges in b^* only;

(iii) any edge in g^* forms a circuit (cutset) of G with edges in g only.

In proving results we occasionally refer to the following proposition which is a more generalized version of a well-known lemma of Tutte:

Proposition 3.8 Given an arbitrary partition $(F, E \setminus F = F^*)$, of the edge set E of a graph, let $e' \in F^*$ form a circuit (cutset) C (S) with edges in F only. Let e be an edge in $C \setminus \{e'\}$ $(S \setminus \{e'\})$. Then every cutset (circuit) that e forms with edges in F^* only, contains e'.

Proof The proof is by contradiction. Suppose that there is a cutset S_e (circuit C_e) that e forms with edges in A^* only, that does not contain e'. Then the intersection of C (S) and S_e (C_e) will contain exactly one edge, (the edge e) which contradicts the Orthogonality Theorem. □

3.3 Transitions from one basoid to another

In this section we present three propositions concerned with transforming one basoid into another. Given a basoid b of a graph G, denote by $Q(b)$ the set of edges in b^* that belong to circuits made of edges in b^* only, and denote by $P(b)$ the set of edges in b^* that belong to cutsets made of edges in b^* only. According to Proposition 3.6 $P(b) \cap Q(b)$ is empty. Thus the pair $(Q(b), P(b))$ generates a partition of b^* into three classes $Q(b)$, $P(b)$ and $b^* \setminus Q(d) \cup P(d)$. Every edge in $Q(b)$ forms a unique circuit with edges in b only and does not form a cutset with edges in b only, every edge in $P(b)$ forms a unique cutset with edges in b only and does not form a circuit with edges in b only and every edge in $b^* \setminus Q(d) \cup P(d)$ forms both a unique circuit and a unique cutset with edges in b only. Also, each edge in b forms both a circuit and a cutset (not necessarily unique) with edges in b^* only.

Proposition 3.9 Given a basoid b of a graph G, let $e' \in b^*$ form both a circuit C and a cutset S with edges in b only. Then, for every $e \in (C \cap S) \setminus \{e'\}$, $b' = (b \setminus \{e\}) \cup \{e'\}$ is also a basoid and the relations $Q(b') = Q(b)$ and $P(b') = P(b)$ hold.

Proof Clearly, the set $(C \cap S) \setminus \{e'\}$ is nonempty because otherwise $|C \cap S| = 1$, which contradicts the Orthogonality Theorem. On the other hand, C is the only circuit made of edges in $b \cup \{e'\}$ only and S is the only cutset made of edges in $b \cup \{e'\}$ only. Choose $e \in (C \cap S)$. After interchanging e and e' the resulting subset $b' = (b \setminus \{e\}) \cup \{e'\}$ is still double independent. Since

e' belongs neither to $Q(b)$ nor to $P(b)$, it follows that $Q(b') \supseteq Q(b)$ and $P(b') \supseteq P(b)$. On the other hand, according to Proposition 3.8, every circuit and every cutset that e forms with edges in b^* only contains e', and hence the edge e forms neither a circuit nor a cutset with edges in $b^* \setminus \{e'\}$ only. Consequently, $Q(b) \supseteq Q(b')$ and $P(b) \supseteq P(b')$ hold.

Thus, $Q(b') \supseteq Q(b)$ and $P(b') \supseteq P(b)$ and at the same time $Q(b) \supseteq Q(b')$ and $P(b) \supseteq P(b')$, which imply $Q(b) = Q(b')$ and $P(b) = P(b')$. The circuits and cutsets, made of edges in b^* only, are disjoint (because b is a basoid). Consequently the circuits and cutsets made of edges in $(b')^*$ only, are also disjoint. Therefore, according to Proposition 3.6, $b' = (b \setminus \{e\}) \cup \{e'\}$ is a basoid. $\square$

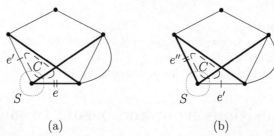

Figure 3.4. Illustrating transition of one basoid to another which preserves cardinality and which keeps all circuits and cutsets made of edges in their complements.

In the graph of Figure 3.4(a), the bold edges indicate a basoid b with cardinality 3. The dotted line indicates a cutset S and the dashed line indicates a circuit C that the edge $e' \in b^*$ forms with edges in b only. According to Proposition 3.9, because $e \in (C \cap S) \setminus \{e'\}$, after exchanging e and e' we obtain another basoid b' with the same cardinality. This is shown in Figure 3.4(b). Conversely, if we start with the basoid b' of Figure 3.4(b), we can obtain the basoid of Figure 3.4(a). Notice that $Q(b') = Q(b)$ and $P(b') = P(b)$.

Proposition 3.10 *Given a basoid b of a graph G, let $x \in Q(b)$ and $y \in P(b)$. Let C be a circuit that x forms with edges in b only and let S be a cutset that y forms with edges in b only. Suppose $C \cap S$ is nonempty. Then for every $e \in C \cap S$, $b' = (b \setminus \{e\}) \cup \{x, y\}$ is a basoid and the relations $Q(b') \subset Q(b)$ and $P(b') \subset P(b)$ hold.*

Proof As x forms a unique circuit C with edges in b only and y forms a unique cutset S with edges in b only, C is the only circuit in $b \cup \{x, y\}$ and S is the only cutset in $b \cup \{x, y\}$. We assume, $C \cap S$ is nonempty. Let $e \in C \cap S$. Then the subset $b' = (b \setminus \{e\}) \cup \{x, y\}$ is obviously double independent. Since, S is the cutset that $e \notin b'$ forms with edges in b' only and C is the circuit that $e \notin b'$ forms with edges in b' only, according to the Painting Theorem, e belongs neither to a circuit nor to a cutset made of edges in $(b')^*$ only. Therefore every circuit (cutset) that entirely belongs to $(b')^*$, belongs also to b^*, that is, $Q(b') \subseteq Q(b)$ and $P(b') \subseteq P(b)$. On the other hand no circuit made of edges in b^* that contains x belongs to $(b')^*$ and no cutset made of edges in b^* only that contains y belongs to $(b')^*$.

Consequently, $Q(b') \subset Q(b)$ and $P(b') \subset P(b)$. Since every circuit in b^* is disjoint to every cutset in b^* it follows that the same property holds for $(b')^*$. Hence, according to Proposition 3.6, b' is a basoid. □

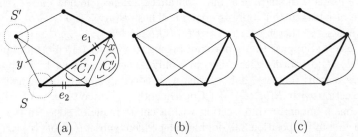

(a)   (b)   (c)

Figure 3.5. Illustrating the transition of one basoid to another which increases the basoid's cardinality by one.

In the graph of Figure 3.5(a), bold edges indicate a basoid with cardinality 3. The edge $x \in b^*$ defines a circuit C with respect to b and at the same time belongs to a circuit C' made of edges in b^* only. The edge $y \in b^*$ defines a cutset S with respect to b and at the same time belongs to a cutset S' made of edges in b^* only. Notice that $C \cap S$ is nonempty. Then, according to Proposition 3.10, for every $e \in C \cap S$, the edge subset $b' = (b \setminus \{e\}) \cup \{x, y\}$ is a basoid. In this example $\{e_1, e_2\} = C \cap S$, and hence there are two transitions from the basoid of Figure 3.5(a), which is with cardinality 3, to the basoids with cardinality 4, as Figures 3.5(b) and 3.5(c), show.

The following proposition strengthens Proposition 3.10 and provides a basis for augmenting basoids in turn.

Proposition 3.11 *Given a basoid b of a graph G, let $e' \in Q(b)$ ($e' \in P(b)$) and let $e \in b$ be an edge that belongs to the unique circuit (cutset) that e' forms with edges in b only.*

(a) *If e does not belong to a cutset (circuit) that an edge in $P(b)$ ($Q(b)$) forms with edges in b only, then $b' = (b \setminus \{e\}) \cup \{e'\} = b \oplus \{e, e'\}$ is a basoid and $P(b') = P(b)$ ($Q(b') = Q(b)$).*

(b) *If e belongs to a cutset (circuit) that an edge $e'' \in P(b)$ ($e' \in Q(b)$) forms with edges in b only then, $b'' = (b \setminus \{e\}) \cup \{e', e''\} = b \oplus \{e, e', e''\}$ is a basoid and $P(b') \subset P(b)$ ($Q(b') \subset Q(b)$).*

Proof Let C' (S') be a circuit (cutset) made of edges in b^* only, that contains the edge $e' \in Q(b)$ $(e' \in P(b))$. According to the Painting Theorem, e' does not form a cutset (circuit) with edges in b only. Hence $b \cup \{e'\}$ is cutset-less (circuit-less). On the other hand, b is basoid and hence e' necessarily forms a circuit (cutset) C (S) with edges in b only. As the unique circuit (cutset) that e' forms with edges in b only, C (S) is the only circuit (cutset) in $b \cup \{e'\}$. Let $e \in C \setminus \{e'\}$ $(e \in S \setminus \{e'\})$. Then the edge subset $b' = (b \setminus \{e\}) \cup \{e'\} = b \oplus \{e, e'\}$ is obviously a double independent edge subset of G. Since e' does not belong to $P(b)$ $(Q(b))$, it follows that every cutset (circuit) made of edges in b^* only, belongs entirely to $(b')^*$. That is, $P(b') \subseteq P(b)$ $(Q(b') \subseteq Q(b))$ holds. On the other hand, according to Proposition 3.8, each cutset (circuit) that e forms with edges in b^* only contains e'. Hence, no new cutset (circuit) made of edges in $(b')^*$ only is created. Thus, $P(b) \subseteq P(b')$ $(Q(b) \subseteq Q(b'))$ also holds and therefore, $P(b') = P(b)$ $(Q(b') = Q(b))$.

Case (a) Suppose that e does not belong to a cutset (circuit) that an edge in $P(b)$ $(Q(b))$ forms with edges in b only. Then, according to Proposition 3.8, no circuit (cutset) that e forms with edges in b^* only has a nonempty intersection with $P(b)$ $(Q(b))$. Consequently, no circuit made of edges in $(b')^*$ has a nonempty intersection with a cutset made of edges in $(b')^*$ and, according to Proposition 3.6, double independent subset $b' = (b \setminus \{e\}) \cup \{e'\} = b \oplus \{e, e'\}$ is a basoid.

Case (b) Suppose now that, unlike case (a), e belongs to a cutset (circuit) that an edge $e'' \in P(b)$ $(e' \in Q(b))$ forms with edges in b only. Then, according to Proposition 3.10, double independent subset $b' = (b \setminus \{e\}) \cup \{e', e''\} = b \oplus \{e, e', e''\}$ is a basoid and $Q(b') \subset Q(b)$ $(P(b') \subset P(b))$. $\square$

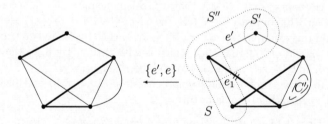

Figure 3.6. Illustrating a basoid transition which preserves the basoid's cardinality.

Figure 3.6 shows a transition from a basoid b with cardinality 3 to another basoid b' with the same cardinality, choosing an edge e' in $P(b)$ and using cutsets. Figure 3.7 shows a transition from the same basoid b with cardinality 3 to another basoid b' with cardinality 4, choosing an edge e' in $Q(b)$ and using circuits. Both transitions are performed according to Proposition 3.11.

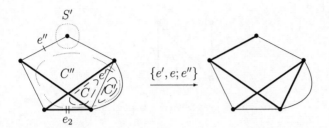

Figure 3.7. Illustrating a basoid transition which augments the basoid's cardinality.

3.4 Minor with respect to a basoid

An edge subset M (respectively, N) is said to be a *minor* (*cominor*) of a graph G with respect to a basoid b if the following two conditions are fulfilled:

  (i) $Q(b) \subseteq M$ ($P(b) \subseteq N$).

  (ii) $sp_c(b \cap M) = M = sp_c(b^* \cap M)$ ($sp_s(b \cap N) = N = sp_s(b^* \cap N)$).

Lemma 3.12 *Let M (N) be an arbitrary edge subset of G and let b be a basoid of G. If the relation $sp_c(b \cap M) \supseteq M$ ($sp_s(b \cap N) \supseteq N$) holds then $P(b) \cap M$ ($Q(b) \cap N$) is empty.*

Proof An edge in b^* belongs to $sp_c(b \cap M)$ ($sp_s(b \cap N)$) iff it forms a circuit (cutset) with edges in $b \cap M$ ($b \cap N$). Since no edge in $P(b)$ ($Q(b)$) forms a circuit (cutset) with edges in b only, it follows that $P(b) \cap sp_c(b \cap M)$ ($Q(b) \cap sp_s(b \cap N)$) is empty. But $sp_c(b \cap M) \supseteq M$ ($sp_s(b \cap N) \supseteq N$) and hence $P(b) \cap M$ ($Q(b) \cap N$) is also empty. □

Notice that if M (N) is a minor (cominor) of a graph G with respect to a basoid b then $sp_c(b \cap M) = M$ ($sp_s(b \cap N) = N$) holds and hence the relation $sp_c(b \cap M) \supseteq M$ ($sp_s(b \cap N) \supseteq N$) immediately follows. Therefore, according to Lemma 3.12, $P(b) \cap M = \emptyset$ ($Q(b) \cap N = \emptyset$) holds.

In the graph of Figure 3.8 bold edges indicate a basoid $b = \{b, c, e\}$ for which $Q(b) = \{g, f\}$ and $P(b) = \{a, h\}$. The edge subsets $\{e, f, g\}$ and $\{c, d, e, f, g\}$ are minors with respect to the basoid $\{b, c, e\}$. Also, the edge subsets $\{a, b, h\}$ and $\{a, b, c, d, h\}$ are cominors with respect to the same basoid.

Lemma 3.13 *Let b be a basoid of a graph and let K be an edge subset of G such that $Q(b) \subseteq K$ and $P(b) \subseteq K^*$. Then the following two equivalencies are true:*

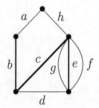

Figure 3.8. A graph with a basoid indicated.

(E_1): $sp_c(b \cap K) \supseteq K$ iff $sp_s(b^* \cap K^*) \supseteq K^*$
and
(E_2): $sp_c(b^* \cap K) \supseteq K$ iff $sp_s(b \cap K^*) \supseteq K^*$.

Proof (E_1): The proof is by contradiction. Suppose that $sp_s(b^* \cap K^*) \supseteq K^*$ is true and $sp_c(b \cap K) \supseteq K$ is not. Then there is an edge $y_1 \in b^* \cap K$ such that the circuit C_1 that y_1 forms with edges in b only, contains $x_1 \in b \cap K^*$. Since, according to Proposition 3.4, part (ii), every edge in b forms a cutset with edges in b^* only, so x_1 does. Denote by S_1 a cutset that x_1 forms with edges in b^* only. According to Proposition 3.8 applied to the partition (b, b^*), if $x_1 \in K^* \cap b$ belongs to the circuit C_1, then y_1 belongs to a cutset S_1. Consequently, $sp_s(b^* \cap K^*) \supseteq K^*$, which is a contradiction.

Conversely, suppose that $sp_c(b \cap K) \supseteq K$ is true and $sp_s(b^* \cap K^*) \supseteq K^*$ is not. Then there is an edge $x_2 \in b \cap K^*$ such that a cutset S_2 that x_2 forms with edges in b^* only, contains $y_2 \in b^* \cap K$. Since $P(b) \cap K$ is empty, every edge in $b^* \cap K$ forms a circuit with edges in b only. Therefore, y_2 forms a circuit C_2 with edges in b only. According to Proposition 3.8 applied to the partition (b, b^*), if $y_2 \in b^* \cap K$ belongs to S_2, then, x_2 belongs to a circuit C_2. Consequently, $sp_c(b \cap K) \supseteq K$, a contradiction.

(E_2): By interchanging the roles of K and K^*, circuits and cutsets, as well $sp_c(\cdot)$ and $sp_s(\cdot)$, we can prove that $sp_c(b^* \cap K) \supseteq K$ iff $sp_s(b \cap K^*) \supseteq K^*$. $\qquad \square$

Proposition 3.14 *Let b be a basoid of a graph and let M be an edge subset of G. Then the following statements are equivalent:*

(1) *M is a minor of G with respect to b.*

(1') *$sp_c(b \cap M) \supseteq M \subseteq sp_c(b^* \cap M)$ and $Q(b) \subseteq M$.*

(2) *M^* is a cominor of G with respect to b.*

(2') *$sp_s(b \cap M^*) \supseteq M^* \subseteq sp_s(b^* \cap M^*)$ and $P(b) \subseteq M^*$.*

(3) *$sp_c(b \cap M) = M$ and $sp_s(b \cap M^*) = M^*$.*

(3') *$sp_c(b \cap M) \supseteq M$ and $sp_s(b \cap M^*) \supseteq M^*$.*

(3'') *$M \cap b$ is a maximal circuit-less subset of M and $M^* \cap b$ is a maximal cutset-less subset of M^*.*

Proof (1) $\Rightarrow$ (1') Obvious.

(1') $\Rightarrow$ (1) Suppose (1') holds. According to Lemma 3.12, $sp_c(b \cap M) \supseteq M$ implies that $P(b) \cap M$ is empty. To prove that M is a minor of G with respect to b it is enough to prove that $sp_c(b \cap M) \setminus M$ and $sp_c(b^* \cap M) \setminus M$ are empty. The proof is by contradiction. Suppose there is an edge $x \in (sp_c(b \cap M) \setminus M)$. Clearly, $sp_c(b \cap M) \setminus M$ is a proper subset of $b^* \cap M^*$ and x forms a circuit C_x with edges in $b \cap M$ only. According to Lemma 3.13, $sp_c(b^* \cap M) \supseteq M$ implies $sp_s(b \cap M^*) \supseteq M^*$, and therefore x also forms a cutset S_x with edges in $b \cap M^*$ only. But $C_x \cap S_x = \{x\}$, which contradicts the Orthogonality Theorem. Suppose now that there is an edge $y \in (sp_c(b^* \cap M) \setminus M)$. Clearly, $sp_c(b \cap M) \setminus M$ is a proper subset of $b \cap M^*$ and y forms a circuit C_y with edges in $b^* \cap M$ only. According to Lemma 3.13, $sp_c(b \cap M) \supseteq M$ implies $sp_s(b^* \cap M^*) \supseteq M^*$ and therefore y also forms a cutset S_y with edges in $b^* \cap M^*$ only. But $C_y \cap S_y = \{y\}$, which contradicts the Orthogonality Theorem.

(2) $\Rightarrow$ (2') Obvious.

(2') $\Rightarrow$ (2) Dual to the proof of (1') $\Rightarrow$ (1).

(3) $\Rightarrow$ (3') Obvious.

(3') $\Rightarrow$ (3) Suppose that (3') holds. According to Lemma 3.12, (3') implies $P(b) \cap M = \emptyset$ and $Q(b) \cap M^* = \emptyset$ and hence $Q(b) \subseteq M$ and $P(b) \subseteq M^*$. Suppose there is an edge $x \in (sp_c(b \cap M) \setminus M)$. Clearly, $sp_c(b \cap M) \setminus M$ is a proper subset of $b^* \cap M^*$ and x forms a circuit C_x with edges in $b \cap M$ only. Because $sp_s(b \cap M^*) \supseteq M^*$, x also forms a cutset S_x with edges in $b \cap M^*$ only. But $C_x \cap S_x = \{x\}$, which contradicts the Orthogonality Theorem. Dually, suppose there is an edge $y \in (sp_s(b \cap M^*) \setminus M^*)$. Clearly, y forms a cutset S_y with edges in $b \cap M^*$ only. Since, $sp_s(b \cap M^*) \setminus M^*$ is a proper subset of $b^* \cap M$ and because $sp_c(b \cap M) \supseteq M$, y also forms a circuit C_y with edges in $b \cap M$ only. But $C_y \cap S_y = \{y\}$, which contradicts the Orthogonality Theorem.

(3'') $\Leftrightarrow$ (3') Obvious.

(1') $\Leftrightarrow$ (3') According to Lemma 3.12, $sp_c(b \cap M^*) \supseteq M^*$ implies $Q(b) \subseteq M$ and, according to Lemma 3.13, $sp_c(b^* \cap M) \supseteq M$ holds iff $sp_s(b \cap M^*) \supseteq M^*$.

(2') $\Leftrightarrow$ (3') According to Lemma 3.12, $sp_s(b \cap M^*) \supseteq M^*$ implies $P(b) \subseteq M^*$ and, according to Lemma 3.13, $sp_c(b \cap M) \supseteq M$ holds iff $sp_s(b^* \cap M^*) \supseteq M^*$. □

Using dual arguments we can prove also the following proposition.

Dual of Proposition 3.14 *Let b be a basoid of a graph G and let N be an edge subset of G. Then the following statements are equivalent:*

(1) *N is a cominor of G with respect to b.*

(1') *$sp_s(b \cap N) \supseteq N \subseteq sp_s(b^* \cap N)$ and $P(b) \subseteq N$.*

(2) *N^* is a minor of G with respect to b.*

(2') *$sp_c(b \cap N^*) \supseteq N^* \subseteq sp_c(b^* \cap N^*)$ and $Q(b) \subseteq N^*$.*

(3) $sp_s(b \cap N) = N$ and $sp_c(b \cap N^*) = N^*$.

(3') $sp_s(b \cap N) \supseteq N$ and $sp_c(b \cap N^*) \supseteq N^*$.

(3'') $N \cap b$ is a maximal cutset-less subset of N and $N^* \cap b$ is a maximal circuit-less subset of N^*.

3.5 Principal sequence

Given a basoid b of a graph G, let $A_Q^0 = Q(b)$. Denote by X_Q^1 the union of all circuits that edges in A_Q^0 form with edges in b only and denote by Y_Q^1 the union of all circuits that edges in $X_Q^1 \cap b$ form with edges in b^* only. Inductively, for $k \geq 1$, denote by X_Q^{k+1} the union of all circuits that edges in $Y_Q^k \cap b^*$ form with edges in b only, and denote by Y_Q^{k+1} the union of all circuits that edges in $X_Q^{k+1} \cap b$ form with edges in b^* only. Because each edge in b forms a circuit with edges in b^* only, it is clear that $X_Q^{k+1} \cap b = Y_Q^{k+1} \cap b$. Also, if the intersection $P(b) \cap Y_Q^k$ is empty, then $X_Q^{k+1} \cap b^* = Y_Q^k \cap b$. This is because every edge in $b^* \setminus P(b)$ forms a circuit with edges in b only. Now define the sequence $\{A_Q^k, \ k \geq 0 \ \}$ by the recurrence relation $A_Q^{k+1} = A_Q^k \cup Y_Q^{k+1}$. We call this a *principal sequence* of edge subsets associated with a basoid b. By construction, $A_Q^k \subseteq A_Q^{k+1}$, for all $k \geq 0$. Because of the finiteness of the edge set of G, there must be a smallest integer $r > 0$, such that $A_Q^r = A_Q^k$, for all $k > r$. For such an integer r we write $A_Q^r = A_Q(b)$.

By construction it is clear that any edge in $(Y_Q^{k+1} \setminus Y_Q^k) \cap b^*$ belongs to a circuit that an edge in $(Y_Q^{k+1} \setminus Y_Q^k) \cap b$ form with edges in b^* only. Also, any edge in $(Y_Q^{k+1} \setminus Y_Q^k) \cap b$ belongs to a circuit that an edge in $(Y_Q^k \setminus Y_Q^{k-1}) \cap b^*$ form with edges in b only. But $Y_Q^{k+1} \setminus Y_Q^k = A_Q^{k+1} \setminus A_Q^k$ and hence, *any edge in $(A_Q^{k+1} \setminus A_Q^k) \cap b^*$ belongs to a circuit that an edge in $(A_Q^{k+1} \setminus A_Q^k) \cap b$ forms with edges in b^* only. Also, any edge in $A_Q^{k+1} \setminus A_Q^k \cap b$ belongs to a circuit that an edge in $A_Q^k \setminus A_Q^{k-1} \cap b^*$ forms with edges in b only.* According to Proposition 3.8, it follows that *the cutset that any edge in $(A_Q^{k+1} \setminus A_Q^k) \cap b^*$ forms with edges in b only has a nonempty intersection with $(A_Q^{k+1} \setminus A_Q^k) \cap b$. Also, the cutset that any edge in $(A_Q^{k+1} \setminus A_Q^k) \cap b$ forms with edges in b^* only has a nonempty intersection with $(A_Q^k \setminus A_Q^{k-1}) \cap b^*$.*

Lemma 3.15 *Let $A_Q(b)$ be the largest subset of a principal sequence associated with a basoid b. Then the following hold: (a) $A_Q(b) \supseteq Q(b)$ and (b) if $P(b) \cap A_Q(b)$ is empty, then $sp_c(b \cap A_Q(b)) \supseteq A_Q(b) \subseteq sp_c(b^* \cap A_Q(b))$.*

Proof (a) follows immediately from the construction of $A_Q(b)$. If $P(b) \cap A_Q(b)$ is empty then every edge in $A_Q(b) \setminus b$ forms a circuit with edges in $A_Q(b) \cap b$ and every edge in $A_Q(b) \setminus b^*$ forms a circuit with edges in $A_Q(b) \cap b^*$. Consequently, $sp_c(b \cap A_Q(b)) \supseteq A_Q(b) \subseteq sp_c(b^* \cap A_Q(b))$, that is, (b) holds. $\square$

In dual manner, for a given basoid b, we can construct the sequence $B_P^0, B_P^1, \ldots, B_P^k, \ldots$ which we call a *principal cosequence* associated with a basoid b of a graph G. Let $B_P(b)$ be the largest subset of a principal cosequence associated with a basoid b.

From the construction of $B_P(b)$ it follows that the Dual of Lemma 3.15, also holds:

Dual of Lemma 3.15 *Let $B_P(b)$ be the largest subset of a principal cosequence associated with a basoid b. Then the following hold:* (a) $B_P(b) \supseteq P(b)$ *and* (b) *if* $Q(b) \cap B_P(b)$ *is empty, then* $sp_s(b \cap B_P(b)) \supseteq B_P(b) \subseteq sp_s(b^* \cap B_P(b))$.

Proposition 3.16 *Let $A_Q(b)$ be the largest subset of a principal sequence associated with a basoid b. Then $A_Q(b)$ is a minor of G with respect to b iff $A_Q(b) \cap P(b)$ is empty.*

Proof Suppose that $A_Q(b)$ is a minor of G with respect to b. Then, according to Proposition 3.14, $A_Q(b)^*$ is a cominor and hence, $P(b) \subseteq A_Q(b)^*$. Therefore, $A_Q(b) \cap P(b)$ is empty.

Conversely, if the intersection $A_Q(b) \cap P(b)$ is empty, then, according to Lemma 3.15, $A_Q(b)$ is a minor. □

Dual of Proposition 3.16 *Let $B_P(b)$ be the largest subset of a principal cosequence associated with a basoid b. Then $B_P(b)$ is a cominor of G with respect to b iff $B_Q(b) \cap Q(b)$ is empty.*

Proposition 3.17 *Let b be a dyad and let $A_Q(b)$ be the largest subset of a principal sequence associated with b. Then $A_Q(b) \cap P(b)$ is empty.*

Proof The proof is by contradiction. Assume that $A_Q(b) \cap P(b)$ is not empty. From the construction of the sequence $A_Q^0 = Q(b), A_Q^1, \ldots, A_Q^k, \ldots$ it follows that there is a positive integer s such that the intersection $P(b) \cap A_Q^s$ is empty and $P(b) \cap A_Q^{s+1}$ is not. Let $\hat{e}_{s+1} \in A_Q^{s+1} \cap P(b)$. By backtracking from $\hat{e}_{s+1}$ we can construct a sequence of edges $\hat{e}_0, e_0, \hat{e}_1, e_1, \ldots, \hat{e}_s, e_s, \hat{e}_{s+1}$ associated with a base b such that the following conditions hold:

(o) $\hat{e}_0 \in A_Q^0 = Q(b)$ and for $0 \leq k \leq s$, $\hat{e}_{k+1} \in (A_Q^{k+1} \setminus A_Q^k) \cap b^*$ and $e_k \in (A_Q^{k+1} \setminus A_Q^k) \cap b$.
and

(i) $\hat{e}_0 \in \hat{C}_0$ and for $0 \leq k \leq s$, $e_k \in C_k \setminus \{\hat{e}_k\}$ and $\hat{e}_{k+1} \in \hat{C}_{k+1} \setminus \{e_k\}$, where, $\hat{C}_0$ is a circuit that $\hat{e}_0$ forms with edges in b^* only and, for $0 \leq k \leq s$, C_k is a circuit that $\hat{e}_k$ forms with edges in b only and $\hat{C}_{k+1}$ is a circuit that e_k forms with edges in b^* only.

It is clear by construction of the sequence $A_Q^0 = Q(b), A_Q^1, \ldots, A_Q^k, \ldots$ that the following conditions also hold:

(ii) $\hat{e}_0 \in Q(b)$,

(iii) $\hat{e}_{s+1} \in P(b)$

(iv) for $0 \leq k \leq s$, $\hat{C}_k \cap P(b) = \emptyset$,

(v) for $0 \leq p < k \leq s$, $e_k \notin C_p$, and $\hat{e}_k \notin \hat{C}_p$

We consider two cases: $s = 0$ and $s > 0$. First suppose that $s = 0$. Then, the above sequence reduces to $\{\hat{e}_0, e_0, \hat{e}_1\}$ and conditions (i)–(v) reduce to $\hat{e}_0 \in \hat{C}_0 \subseteq Q(b)$, $e_0 \in C_0 \setminus \{\hat{e}_0\}$ and $\hat{e}_1 \in \hat{C}_1 \setminus \{e_0\} \subseteq P(b)$. According to Proposition 3.8 e_0 belongs to a cutset $\hat{S}_1$ that $\hat{e}_1 \in P(b)$ forms with edges in b only. Thus, $e_0 \in \hat{S}_1 \cap C_0$ and hence, according to Proposition 3.11, part (b), $b_1 = (b \setminus \{e_0\}) \cup \{\hat{e}_0, \hat{e}_1\} = b \oplus \{\hat{e}_0, e_0\}$ is an augmented basoid. This contradicts our assumption that b is of maximal cardinality.

Suppose $s > 0$. Then, according to condition (iv), any circuit $\hat{C}_1$ that e_0 forms with edges in b^* only, has empty an intersection with $P(b)$. According to Proposition 3.8, e_0 does not belong to a cutset that an edge in $P(b)$ forms with edges in b only. Then, according to Proposition 3.11, part (a), $b_1 = (b \setminus \{e_0\}) \cup \{\hat{e}_0\}$ is a basoid with the same cardinality as b and $P(b_1) = P(b)$. The sequence $\hat{e}_1, e_1, \ldots, \hat{e}_s, e_s, \hat{e}_{s+1}$ obviously satisfies conditions (o)–(v) with respect to a basoid $b_1 = b \oplus \{\hat{e}_0, e_0\}$. This is in terms of circuits $\hat{C}_k^1$ and C_k^1 defined for all $1 \leq k(\leq s)$ as follows: $\hat{C}_k^1 = \hat{C}_k \oplus \hat{C}_0$ if $\hat{C}_k$ contains $\hat{e}_0$, and $\hat{C}_k^1 = \hat{C}_k$ otherwise; $C_k^1 = C_k \oplus C_0$ if C_k contains e_0, and $C_k^1 = C_k$ otherwise. Since, according to condition (v), for every $1 \leq k(\leq s)$, $\hat{e}_{k+1}$ does not belong to the circuit $\hat{C}_1$, we conclude that for every $1 \leq k(\leq s)$, $\hat{C}_{k+1}^1$ still contains $\hat{e}_{k+1}$. Also, since for $1 \leq k(\leq s)$, e_{k+1} does not belong to the circuit C_1, we conclude that for every $1 \leq k(\leq s)$, C_{k+1}^1 still contains e_{k+1}.

Inductively, for some $q < s$, suppose that sequence $\hat{e}_q, e_q, \hat{e}_{q+1}, e_{q+1}, \ldots$, $\hat{e}_s, e_s, \hat{e}_{s+1}$, satisfies conditions (o)–(v) with respect to a basoid b_q in terms of circuits $\hat{C}_k^q$ and C_k^q, $q \leq k(\leq s)$. Then, using similar arguments to those above, it is easy to see that the sequence $\hat{e}_{q+1}, e_{q+1}, \ldots, \hat{e}_s, e_s, \hat{e}_{s+1}$ satisfies conditions (o)–(v) with respect to a basoid $b_{q+1} = b_q \oplus \{\hat{e}_q, e_q\}$. This is in terms of circuits $\hat{C}_k^{q+1}$ and C_k^{q+1} defined for all $q + 1 \leq k(\leq s)$ as follows: $\hat{C}_k^{q+1} = \hat{C}_k^q \oplus \hat{C}_q^q$ if $\hat{C}_k^q$ contains $\hat{e}_q$, and $\hat{C}_k^{q+1} = \hat{C}_k^q$ otherwise; $C_k^{q+1} = C_k^q \oplus C_q^q$ if C_k^q contains e_q, and $C_k^{q+1} = C_k^q$ otherwise. Since, according to condition (v), for every $q+1 \leq k(\leq s)$, $\hat{e}_{k+1}$ does not belong to the circuit $\hat{C}_{q+1}^q$, we conclude that for every $q + 1 \leq k(\leq s)$, $\hat{C}_{k+1}^{q+1}$ still contains $\hat{e}_{k+1}$. Also, since, for $q + 1 \leq k(\leq s)$ e_{k+1} does not belong to the circuit C_{q+1}^q, we conclude that for every $q+1 \leq k(\leq s)$, C_{k+1}^{q+1} still contains e_{k+1}. Thus, for all $q < s$, $b_{q+1} = b_q \oplus \{\hat{e}_q, e_q\}$ is a basoid with the same cardinality as b and $P(b_{q+1}) = P(b)$. For $q = s$, the triple $\hat{e}_s, e_s, \hat{e}_{s+1}$ is a sequence with respect to b_s. Using arguments similar to those for $s = 0$, we can show that $b_{s+1} = b_s \setminus \{e_s\} \cup \{\hat{e}_s, \hat{e}_{s+1}\} = b_s \oplus \{\hat{e}_s, e_s, \hat{e}_{s+1}\}$ is a basoid with cardinality one greater than b. This contradicts the assumption that b is a dyad.   □

Lemma 3.18 *Let M be a minor of a basoid b of a graph G. Then for an arbitrary basoid b_1 of G the following relations hold: $|b_1| = |b_1 \cap M| + |b_1 \cap M^*| \leq |b \cap M| + |b \cap M^*| = |b|$.*

Proof Let M be a minor of G with respect to a basoid b. Then, according to Proposition 3.14, $b \cap M$ is a maximal circuit-less subset of M and $b \cap M^*$ is a maximal cutset-less subset of M. Let b_1 be an arbitrary basoid of G. Since $b \cap M$ is a maximal circuit-less subset of M and because $b_1 \cap M$ is a circuit-less subset of M, it follows that, $|b_1 \cap M| \leq |b \cap M|$. Dually, since $b \cap M^*$ is a maximal cutset-less subset of M^*, and since $b_1 \cap M^*$ is a cutset-less subset of M^*, it follows that $|b_1 \cap M^*| \leq |b \cap M^*|$. On the other hand, from the obvious fact that (M, M^*) is a partition of the edge set of G, it follows that $|b| = |b \cap M| + |b \cap M^*|$ and $|b_1| = |b_1 \cap M| + |b_1 \cap M^*|$. Consequently, $|b_1| = |b_1 \cap M| + |b_1 \cap M^*| \leq |b \cap M| + |b \cap M^*| = |b|$. □

Proposition 3.19 *A basoid b of a graph G is a dyad of G iff there is an edge subset K of G such that $K \cap b$ is a maximal circuit-less subset of K and $K^* \cap b$ is a maximal cutset-less subset of K^*.*

Proof Let b be a dyad. Then, according to Proposition 3.17, $A_Q(b) \cap P(b)$ is empty which, according to Proposition 3.16 implies that $K = A_Q(b)$ is a minor with respect to b. According to (1)$\Rightarrow$(4) of Proposition 3.14, we conclude that $K \cap b$ is a maximal circuit-less subset of K and $K^* \cap b$ is a maximal cutset-less subset of K^*.

Conversely, let K be an edge subset of G such that $K \cap b$ is a maximal circuit-less subset of K and $K^* \cap b$ is a maximal cutset-less subset of K^*. Then, according to (4)$\Rightarrow$(1) of Proposition 3.14, K is a minor of b and hence, according to Lemma 3.18, for an arbitrary basoid b_1, the following relations hold: $|b_1| = |b_1 \cap K| + |b_1 \cap K^*| \leq |b \cap K| + |b \cap K^*| = |b|$. But b_1 is an arbitrary basoid of G and therefore, b is a dyad. □

Corollary 3.20 [to Propositions 3.14 and 3.19] *Let b be a basoid of a graph G. Then there is a minor of G with respect to b iff b is a dyad.*

Lemma 3.21 *Let M be a minor of G with respect to basoid b. Then, $A_Q(b) \subseteq M$.*

Proof Let M be a minor of G with respect to basoid b. Then $Q(b) \subseteq M$. Clearly, every edge in $Q(b)$ forms a circuit with edges in b only. Let X_Q^1 be the union of all circuits that edges in $Q(b)$ form with edges in b only. Because $Q(b) \subseteq M$ and $sp_c(b \cap M) = M$ we conclude that $X_Q^1 \subseteq M$. Let Y_Q^1 be the union of all circuits that edges in $X_Q^1 \cap b$ form with edges in b^* only. Because $X_Q^1 \subseteq M$ and $sp_c(b^* \cap M) = M$ we conclude that $Y_Q^1 \subseteq M$. Inductively, for $k \geq 1$, let $Y_Q^k \subseteq M$. Since $M \cap P(b)$ is empty, the intersection $P(b) \cap Y_Q^k$ is also empty, and therefore every edge in $Y_Q^k \cap b^*$ forms a circuit with edges in b only. Let X_Q^{k+1} be the union of all circuits that edges in $Y_Q^k \cap b^*$ form with edges in b only. Because $Y_Q^k \subseteq M$ and $sp_c(b \cap M) = M$ we conclude that $X_Q^{k+1} \subseteq M$. Let Y_Q^{k+1} be the union of all circuits that edges in $X_Q^{k+1} \cap b$ form with edges in b^* only. Because $X_Q^{k+1} \subseteq M$ and $sp_c(b^* \cap M) = M$ we conclude that $Y_Q^{k+1} \subseteq M$. Let, $A_Q^0 = Q(b)$ and let $A_Q^{k+1} = A_Q^k \cup Y_Q^{k+1}$. By construction, $A_Q^k \subseteq A_Q^{k+1}$, for all $k \geq 0$. Because of the finiteness of the edge set of G, there is a smallest integer $s > 0$, such that $A_Q^s = A_Q^{k+1}$, for all $k \geq s$. For such an integer s we denote $A_Q^s = A_Q(b)$. Accordingly, for all $k \geq 0$, $A_Q^k \subseteq M$ and therefore $A_Q(b) \subseteq M$.   □

Dual of Lemma 3.21 Let N be a minor of G with respect to basoid b. Then, $B_P(b) \subseteq N$.

Proposition 3.22 *Given a basoid b, let $A_Q(b)$ be the largest subset of a principal sequence associated with b and let $B_P(b)$ be the largest subset of a principal cosequence associated with b. Then the following statements are equivalent:*

(1) *A basoid b of a graph G is a dyad.*

(2) $A_Q(b) \cap P(b)$ *is empty.*

(3) $B_Q(b) \cap Q(b)$ *is empty.*

(4) $A_Q(b) \cap B_P(b)$ *is empty.*

Proof (1)⇔(2) Suppose b is a dyad of a graph G. Then, according to Proposition 3.19, there is an edge subset K of G such that $K \cap b$ is a maximal circuit-less subset of K and $K^* \cap b$ is a maximal cutset-less subset of K^*. According to Proposition 3.14, K is a minor of G with respect to b and hence, according to Lemma 3.21, $A_Q(b) \subseteq K$. Since $K \cap P(b)$ is empty, it follows that $A_Q(b) \cap P(b)$ is also empty. Thus (1)⇒(2) is true. This may also be proved by reference to Proposition 3.17.

Conversely, if $A_Q(b) \cap P(b)$ is empty then, according to Lemma 3.15, $Q(b) \subseteq A_Q(b)$ and $sp_c(b \cap A_Q(b)) \supseteq A_Q(b) \subseteq sp_c(b^* \cap A_Q(b))$, that is, $A_Q(b)$ is a minor of G with respect to b. According to Proposition 3.14 this implies that $A_Q(b) \cap b$ is a maximal circuit-less subset of $A_Q(b)$ and $A_Q(b)^* \cap b$ is a maximal cutset-less subset of $A_Q(b)^*$. According to Proposition 3.19, this means that b is a dyad.

(1)⇔(3) The proof is dual to (1)⇔(2).

(2)⇔(4) Suppose $A_Q(b) \cap P(b)$ is empty. Then, according to Lemma 3.15, $Q(b) \subseteq A_Q(b)$ and $sp_c(b \cap A_Q(b)) \supseteq A_Q(b) \subseteq sp_c(b^* \cap A_Q(b))$. That is, $A_Q(b)$ is a minor of G with respect to b. It follows from Proposition 3.14 that $A_Q(b)^*$ is a cominor. According to the Dual of Lemma 3.21 this implies that $B_P(b) \subseteq A_Q(b)^*$ and therefore $B_P(b) \cap A_Q(b)$ is empty.

Conversely, suppose that $A_Q(b) \cap B_P(b)$ is empty. Since $P(b) \subseteq B_P(b)$ it immediately follows that $A_Q(b) \cap P(b)$ is also empty.   □

Corollary 3.23 [to Propositions 3.16 and 3.22] *Let b be a basoid of a graph G. Then $A_Q(b)$ is a minor of G with respect to b iff b is a dyad.*

3.6 Principal minor and principal partition

A minor M of G with respect to a dyad b is said to be a *minimal minor* of G with respect to b if no proper subset of M is also a minor of G with respect to b. We will prove that there is a unique edge subset of G which is the minimal minor with respect to every dyad of G. We start with the following lemma.

Lemma 3.24 *Let M' and M'' be two minors of a graph G with respect to a dyad b. Then, $M' \cap M''$ is also a minor of G with respect to b.*

Proof Let M' and M'' be two minors of G with respect to a dyad b and let $\hat{e} \in M' \cap M'' \cap b^*$. Since $sp_c(b \cap M') = M'$ and $sp_c(b \cap M'') = M''$, $\hat{e}$ forms a circuit C'_b with edges in $M' \cap b$ and a circuit C''_b with edges in $(M'' \cap b)$ only. But every edge in b^* forms a unique circuit with edges in b and therefore $C'_b = C''_b \subseteq M' \cap M''$. Consequently, $sp_c(b \cap M' \cap M'') \supseteq M' \cap M''$.

Consider now an edge $e \in M' \cap M'' \cap b$. Since $sp_c(b^* \cap M') = M'$ and $sp_c(b^* \cap M'') = M''$, e forms at least one circuit with edges in $M' \cap b^*$ only and at least one circuit with edges in $M'' \cap b^*$ only. Because a circuit that an edge in b forms with edges in b^* is generally not unique, these two circuits may differ. If they coincide, then clearly both must belong entirely to $M' \cap M''$. Otherwise, if they are distinct, they generally may not entirely belong to $M' \cap M''$. Suppose there is no circuit that e forms with edges in $b^* \cap M' \cap M''$ only. Let C'_{b^*} be a circuit that e forms with edges in $b^* \cap M'$ only and let C''_{b^*} be a circuit that e forms with edges in $b^* \cap M''$ only. Because neither of them entirely belongs to $M' \cap M''$, it follows that C'_{b^*} contains at least one edge $x' \in M' \backslash M''$ and C''_{b^*} contains at least one edge $x'' \in M'' \backslash M'$. Clearly, because $Q(b)$ entirely belongs to the intersection $M' \cap M''$, neither x' nor x'' belong to $Q(b)$. As $x' \in C'_{b^*} \setminus C''_{b^*}$ and $e \in C'_{b^*} \cap C''_{b^*}$, according to the Circuit axiom (see section 3.1), there is a circuit C_{b^*} such that $x' \in C_{b^*} \subseteq (C'_{b^*} \cup C''_{b^*}) \setminus \{e\}$. Because $C'_{b^*} \setminus \{e\}$ and $C''_{b^*} \setminus \{e\}$ both entirely belong to b^* it follows that C_{b^*} also entirely belongs to b^* and consequently belongs to $Q(b)$. Therefore $x' \in Q(b)$, which is a contradiction. Thus the assumption that $x' \in M' \setminus M''$ produces a contradiction. Using the same arguments we can prove that the assumption $x'' \in M'' \setminus M'$ will also produce a contradiction. Thus both C'_{b^*} and C''_{b^*} belong entirely to $M' \cap M''$. Consequently, $sp_c(b^* \cap (M' \cap M'')) \supseteq M' \cap M''$.

Since, $sp_c(b \cap (M' \cap M'')) \supseteq M' \cap M''$ and $sp_c(b^* \cap (M' \cap M'')) \supseteq M' \cap M''$, according to $(1) \Leftrightarrow (1')$ of Proposition 3.14, we conclude that $M' \cap M''$ is a minor of G with respect to b. $\square$

Proposition 3.25 *Let b be a dyad of a graph G. Then, there is a unique minimal minor of G with respect to b.*

Proof Consider the family consisting of all minors of a graph G with respect to a dyad b. According to Corollary 3.23, if b is dyad of G then $A_Q(b)$ is a minor of G with respect to b. Therefore, the family of all the minors of a graph G with respect to a dyad b is not empty. According to Lemma 3.24, the intersection of all the minors is a member of the family and is a proper subset of each member of the family. Hence, it is the unique minimal subset of this family.   □

Proposition 3.26 *Let b be a dyad of a graph G and let $A_Q(b)$ be the largest subset of a principal sequence associated with b. Then, the unique minimal minor of G with respect to b coincides with the set $A_Q(b)$.*

Proof Let M be a minimal minor of a graph G with respect to a dyad b of G. Since b is a dyad, according to Corollary 3.23, $A_Q(b)$ is a minor of G with respect to b. Thus, M and $A_Q(b)$ are both minors of G with respect to b. But, by our assumption, M is the minimal one with respect to b and consequently, $M \subseteq A_Q(b)$. On the other hand, according to Lemma 3.21, $A_Q(b) \subseteq M$. Thus, at the same time, $M \subseteq A_Q(b)$ and $A_Q(b) \subseteq M$, which imply $M = A_Q(b)$.   □

Lemma 3.27 *Let M be a minor of a graph G with respect to a basoid b of G. Then $\operatorname{rank}(M) + \operatorname{corank}(M^*) = |b|$.*

Proof From Proposition 3.14, $sp_c(b \cap M) = M$ and $sp_s(b \cap M^*) = M^*$ and hence, $\operatorname{rank}(M) = |b \cap M|$ and $\operatorname{corank}(M^*) = |b \cap M^*|$. But $|b| = |b \cap M| + |b \cap M^*|$ and finally we obtain $|b| = \operatorname{rank}(M) + \operatorname{corank}(M^*)$.   □

The next proposition strengthens Proposition 3.25.

Proposition 3.28 *There is a unique minimal minor of G with respect to every dyad of G.*

Proof Let M' be a minimal minor of G with respect to a dyad b' of G and let M'' be a minimal minor of G with respect to a dyad b'' of G. Clearly,

$$|b'| = |b''|. \tag{1}$$

Because M' is a minor of b', according to Lemma 3.18

$$|b''| = |b'' \cap M'| + |b'' \cap (M')^*| \leq |b' \cap M'| + |b' \cap (M')^*| = |b'|. \tag{2}$$

From (1) and (2) we conclude that

$$|b'' \cap M'| + |b'' \cap (M')^*| = |b'|. \tag{3}$$

Because $b'' \cap M'$ is circuit-less and $b'' \cap (M')^*$ is cutset-less, it follows that

$$\text{rank}\,(M') \geq |b'' \cap M'|, \tag{4a}$$

$$\text{corank}\,((M')^*) \geq |b'' \cap (M')^*|. \tag{4b}$$

Using (3), (4a) and (4b) we obtain:

$$\text{rank}\,(M') + \text{corank}\,((M')^*) \geq |b'' \cap M'| + |b'' \cap (M')^*| = |b'|.$$

On the other hand, according to Lemma 3.27, because M' is a minor of G with respect to b', $\text{rank}\,(M') + \text{corank}\,((M')^*) = |b'|$ and therefore

$$\text{rank}\,(M') + \text{corank}\,((M')^*) = |b'' \cap M'| + |b'' \cap (M')^*|. \tag{5}$$

From the relations (4a), (4b) and (5) it follows immediately that $\text{rank}\,(M') = |b'' \cap M'|$ and $\text{corank}\,((M')^*) = |b'' \cap (M')^*|$, that is, $b'' \cap M'$ is a maximal circuit-less subset of M' and $b'' \cap (M')^*$ is a maximal cutset-less subset of $(M')^*$. From (1)⇔(4) of Proposition 3.14 this means that M' is a minor of G with respect to b''. But M'' is the unique minimal minor of G with respect to b'' and hence $M'' \subseteq M'$. By repeating the above arguments, with M' and M'' interchanged, we obtain $M' \subseteq M''$. Since $M'' \subseteq M'$ and $M' \subseteq M''$ it follows that $M' = M''$. □

Thus, for any graph G, there is the unique edge subset M_o which is a minimal minor of G with respect to every dyad of G. We call this subset of edges the *principal minor* of G. From Proposition 3.26, to find M_o it is sufficient to find the set $A_Q(b)$ with respect to any dyad b of G.

We could develop the material of this section in an entirely dual manner, substituting cutsets with circuits, $B_P(b)$ with $A_Q(b)$ and cominor with minor. We say that a cominor N of G with respect to a dyad b of G is a *minimal cominor* of G with respect to b if no proper subset of N is also a cominor of G with respect to b. Consequently, we have the following statements:

Dual of Proposition 3.25 *Let b be a dyad of a graph G. Then, there is a unique minimal cominor of G with respect to b.*

Dual of Proposition 3.26 *Let b be a dyad of a graph G and let $B_P(b)$ be the largest subset of a principal cosequence associated with b. Then a unique minimal cominor of G with respect to b coincides with the edge set $B_P(b)$.*

Dual of Proposition 3.28 *There is a unique edge subset of a graph G which is a minimal cominor of G with respect to every dyad of G.*

Thus, for any graph G, there is a unique edge subset N_o which is a minimal cominor of G, with respect to every dyad of G. We call this the *principal cominor* of G. From the dual of Proposition 3.26, to find N_o it is sufficient to find the set $B_P(b)$ with respect to a dyad b of G.

According to Proposition 3.26, for each dyad b of a graph G, $A_Q(b)$ coincides with the principal minor M_o of G and $B_P(b)$ coincides with the principal cominor N_o of G. From Proposition 3.22, $A_Q(b)$ and $B_P(b)$ are disjoint and therefore M_o and N_o are disjoint. Thus the following proposition holds:

Proposition 3.29 *The principal minor and the principal cominor of a graph G are disjoint.*

We are now in a position to define another interesting graph concept called the principal partition of a graph. Given a graph G, let M_o be the principal minor and let N_o be the principal cominor of G. According to Proposition 3.29, $M_o \cap N_o$ is empty and hence M_o and N_o defines a partition $(M_o, N_o, K_o = E \setminus (M_o \cup N_o))$ of the edge set of G. This partition is called the *principal partition* of a graph.

For the graph of Figure 3.9, the bold edges show the principal minor $M_o = \{g, e, f\}$, the dashed edges show the principal cominor $N_o = \{a, b, h\}$. The principal partition is then a partition $(\{g, e, f\}, \{c, d\}, \{a, b, h\})$.

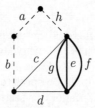

Figure 3.9. A graph with the principal minor indicated with bold edges and the principal co-minor indicated with dashed edges and illustrates the principal partition of the graph.

The principal partition was formally introduced for the first time in the context of tree pairs and then by other authors in more general contexts.

As we shall see in Chapter 5, the concept of principal partition introduced in this chapter with respect to basoids of maximum cardinality (dyads) coincides with the concept of principal partition with respect to pairs of trees of maximum Hamming distance.

3.7   Hybrid rank and basic pairs of subsets

The notion of *hybrid rank* of a graph has been introduced as the minimum rank (K) + corank (K^*) over all edge subsets K. The hybrid rank of a graph has important implications in electrical network theory. In the literature devoted to the analysis of electrical networks, this number is also called the *topological degree of freedom*. We define the hybrid rank of a graph to be the cardinality of a largest both circuit-less and cutset-less subset (dyad) of the graph. This is a natural definition since rank is defined to be the cardinality of a maximal circuit-less subset (tree) and corank is defined to be the cardinality of a maximal cutset-less subset (cotree) of a graph. Because a double independent subset of a graph is both circuit-less and cutset-less, it follows immediately that hybrank $(G) \leq$ rank (G) and hybrank $(G) \leq$ corank (G).

Lemma 3.30 *Let K be an arbitrary edge subset of G and let b be an arbitrary basoid of G. Then, $|b| \leq$ rank $(K) +$ corank (K^*).*

Proof Both $K \cap b$ and $K^* \cap b$ are double independent and therefore $K \cap b$ is circuit-less and $K^* \cap b$ is cutset-less. Hence, $|K \cap b| \leq$ rank (K) and $|K^* \cap b| \leq$ corank (K^*). But $|b| = |b \cap K| + |b \cap K^*|$ and therefore, $|b| \leq$ rank $(K) +$ corank (K^*). □

The following proposition proves that our definition of hybrid rank is equivalent to the original one.

Proposition 3.31 *Let K be an arbitrary edge subset of G. Then* hybrank $(G) = \min_{K \subseteq E}\{$rank $(K) +$ corank $(K^*)\}$.

Proof Let K be an arbitrary edge subset of G and let b be an arbitrary basoid of G. Then, according to Lemma 3.30, $|b| \leq$ rank $(K) +$ corank (K^*). Consequently,

$$\max_b |b| \leq \min_{K \subseteq E}\{\text{rank}(K) + \text{corank}(K^*)\}.$$

Let b' be a dyad of a graph G, that is, let $|b'| = \max_b |b| =$ hybrank (G). Then,

$$\text{hybrank}(G) \leq \min_{K \subseteq E}\{\text{rank}(K) + \text{corank}(K^*)\}. \qquad (1)$$

On the other hand, from Corollary 3.20, there is a minor M of G with respect to b'. This implies (Lemma 3.27) that hybrank $(G) = |b'| =$ rank $(M) +$ corank (M^*). Clearly $\min_{K \subseteq E}\{$rank $(K) +$ corank $(K^*)\} \leq$ rank $(F) +$ corank (F^*) for every edge subset F of G, including $F = M$, and hence:

$$\min_{K \subseteq E}\{\text{rank}(K) + \text{corank}(K^*)\} \leq \text{hybrank}(G). \qquad (2)$$

From (1) and (2) it follows that

$$\text{hybrank}(G) = \min_{K \subseteq E}\{\text{rank}(K) + \text{corank}(K^*)\}.$$

□

Let (τ, μ) be a pair of disjoint edge subsets of E such that τ is circuit-less, μ is cutset-less and $sp_c(\tau) \cup sp_s(\mu) = E$. Such a pair (τ, μ), is called a *basic pair* of subsets. The number $|\tau| + |\mu|$ may differ for different basic pairs of subsets of the network. Basic pairs with a minimum number $|\tau| + |\mu|$ are of particular interest. The following lemma gives a lower bound for $|\tau| + |\mu|$.

Lemma 3.32 hybrank $(G) \leq |\tau| + |\mu|$ *over all basic pairs* (τ, μ) *of G.*

Proof Let (τ, μ) be an arbitrary basic pair of subsets of E and let (E', E'') be a partition satisfying simultaneously the relations: $\tau \subseteq E' \subseteq sp_c(\tau)$, $\mu \subseteq E'' \subseteq sp_s(\mu)$ and $E' \cup E'' = E$. Then, from Proposition 3.31, hybrank $(G) \leq$ rank (E')+corank (E''). But rank $(E') = |\tau|$ and corank $(E'') = |\mu|$ and hence hybrank $(G) \leq |\tau| + |\mu|$, which proves the lemma. $\square$

Proposition 3.33 *Let b be a dyad of a graph G and let M (N) be a minor (cominor) of a graph G with respect to b. Then the pair $(\tau_M = b \cap M, \mu_M = b \cap M^*)$ is a basic pair. Moreover, $|\tau_M| + |\mu_M| \leq |\tau| + |\mu|$ over all basic pairs (τ, μ) of G.*

Proof From Proposition 3.14, if M is a minor of G with respect to a dyad b of G then $sp_c(b \cap M) = M$ and $sp_s(b \cap M^*) = M^*$. Hence, $sp_c(b \cap M) \cup sp_s(b \cap M^*) = M \cup M^* = E$. As b is both circuit-less and cutset-less, every subset of b is also both circuit-less and cutset-less. In particular, $b \cap M$ is circuit-less and $b \cap M^*$ is cutset-less. Clearly, $b \cap M$ and $b \cap M^*$ are disjoint and therefore the pair $(\tau = b \cap M, \mu = b \cap M^*)$ is a basic pair. Because the pair $(\tau_M = b \cap M, \mu_M = b \cap M^*)$ cover b it follows that $|\tau_M| + |\mu_M| = |b|$. On the other hand, according to Lemma 3.30, Proposition 3.31 and Lemma 3.32, $|b| \leq$ hybrank $(G) \leq |\tau| + |\mu|$ over all basic pairs (τ, μ) of G. Hence, $|\tau_M| + |\mu_M| \leq$ hybrank $(G) \leq |\tau| + |\mu|$ over all basic pairs (τ, μ) of G. $\square$

From Proposition 3.33, $|\tau_M| + |\mu_M| \leq |\tau| + |\mu|$ over all basic pairs (τ, μ) of G. The problem of finding a dyad is a special case of the well-known problem of finding a maximum cardinality, maximal subset independent in two matroids. This is a maximum 2-matroid intersection problem. Starting from a graph G and a basoid b of G we can find a dyad b and its partition $(b \cap M, b \cap M^*)$. Besides these basic pairs, there are also several others of minimum cardinality. When $Q(b) = \emptyset$ and $P(b) = \emptyset$, $(b, \emptyset)$ or $(\emptyset, b)$ are basic pair of subsets.

As an immediate consequence of Propositions 3.14 and 3.19, we have the following proposition.

Proposition 3.34 *A basoid b of a graph G is a dyad iff there is a partition (b', b'') of b which is a basic pair of G.*

Proof A basoid b of a graph G is a dyad of G iff there is an edge subset K of G such that $K \cap b$ is a maximal circuit-less subset of K and $K^* \cap b$ is a maximal cutset-less subset of K^* (Proposition 3.19), that is, iff $sp_c(K \cap b) = K$ and $sp_s(K^* \cap b) = K^*$ (Proposition 3.14). Set $K \cap b = b'$ and $K^* \cap b = b''$. Because b' is circuit-less b'' is cutset-less and $sp_c(b') \cup sp_s(b'') = K \cup K^* = E$ we conclude that (b', b'') is a basic pair of G. $\qquad\Box$

For example, bipartitions $(\{b\}, \{c, e\})$ and $(\{b, c\}, \{e\})$ of the dyad $\{b, c, e\}$ of the graph of Figure 3.8, are basic pairs.

3.8 Hybrid analysis of networks

Hybrid network analysis has appeared as a unifying generalization of the well-known mesh and nodal analysis. Various forms of hybrid analysis have been reported in the literature and widely used as topological tools in the formulation of some special type of network equations. In contrast to mesh and nodal analysis, in hybrid analysis the number of equations may vary for the same network depending on the choice of independent variables. This phenomenon, although seemingly an imperfection, provides the possibility of formulating a system of equations, the minimum number of which is less than or equal to the minimum of the numbers of equations for mesh and nodal analysis. This number coincides with the so-called hybrid rank of a graph and has important implications in electrical network theory. In the literature devoted to the analysis of electrical networks, this number is also called the topological degree of freedom. In this section an algorithm for finding a topologically complete set of independent network variables with minimum cardinality, based on 2-matroid intersection, is presented. Discussion of computational cost is also included.

In what follows E denotes the set of ports of a network. We assume that each port in E is resistive, both v-controlled and i-controlled. The galvanic interconnection of ports in a network will be described by a graph associated with the network, which we shall denote by G. Because there is a one-to-one correspondence between the set of network ports and the set of graph edges, E will denote both sets.

In mesh and nodal analysis a partition (E_2, E_1) of E plays a key role. Here E_2 is a maximal circuit-less subset of E (that is, a forest) and $E_1 = E \backslash E_2$ is a maximal cutset-less subset of E (that is, a coforest). On the other hand, hybrid topological analysis is based on a 4-partition $(E_2', E_1', E_2'', E_1'')$ of the set E, associated with a 2-partition (E', E''), which can be obtained according to the following procedure:

Procedure 1 (to find a four partition of E)
input: A bipartition (E', E'') of E.

Step 1: Remove all edges in G that belong to E'' and in the graph obtained find a tree. Denote the tree by E'_2 and the associated cotree by E'_1.
Step 2: Contract all edges in G that belong to E' and in the graph obtained find a tree. Denote the tree by E''_2 and the associated cotree by E''_1.
output: A four partition $(E'_2, E'_1, E''_2, E''_1)$ of E.
end of Procedure 1.

Using standard algorithms, it is easy to encode Procedure 1 so that it runs in $O(\max\{|V|, |E|\})$ time where V is the vertex set and E is the edge set of the graph G, associated with the network.

Obviously, this four partition has the following properties:

(i) the edges of $E'_2 + E''_2$ constitute a tree (E_2) and, the edges of $E'_1 + E''_1$ constitute a cotree (E_1)

(ii) the edges of E'_1 define fundamental loops (f-loops) with edges of E'_2 only and, the edges of E''_2 define fundamental cutsets (f-cutsets) with edges of E''_1 only.

(iii) $E'_1 + E'_2 = E'$ and $E''_1 + E''_2 = E''$.

The four partition $(E'_2, E'_1, E''_2, E''_1)$ and its properties are schematically illustrated in Figure 3.10. Conditions (i) and (ii) describe the topological completeness of the edge partition. By this we mean that all port voltages of E' (that is, v') are determined by port voltages of E'_2 (that is, v'_2) through Kirchhoff's voltage law (KVL), and all port currents of E'' (that is, i'') are determined by port currents of E''_1 (that is, i''_1) through Kirchhoff's current law (KCL). Condition (iii) expresses the compatibility of the topological partition with the way the network ports are controlled.

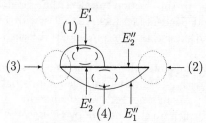

Figure 3.10. Schematical illustration of a four partition $(N'_2, N'_1, N''_2, N''_1)$.

The set of independent variables $\{v'_2, i''_1\}$ of a network, associated with a pair (E'_2, E''_1) is called a *topologically complete set* of variables. The values of these variables can be chosen in an arbitrary manner without violating the topological relations. The cardinalities of topologically complete sets of variables of a network generally may differ with different partitions (E', E'') of the edge set E of a graph. Those with minimal cardinality are of particular interest. Proposition 3.34 states that the minimum number of topologically complete variables of a network is equal to the hybrid rank of the associated graph G.

The set of necessary topological relations (that is, voltage and current Kirchhoff's laws) written on the basis of a partition $(E'_2, E'_1, E''_2, E''_1)$, has the form

$$v'_1 = -P^{\mathrm{T}} v'_2 \tag{1'}$$

$$i''_2 = Q^{\mathrm{T}} i''_1 \tag{2''}$$

$$i'_2 = P i'_1 + B^{\mathrm{T}}_{1''2'} i''_1 \tag{2'}$$

$$v''_1 = -Q v''_2 - B_{1''2'} v'_2, \tag{1''}$$

where (v'_2, i'_2), (v''_2, i''_2), (v'_1, i'_1) and (v''_1, i''_1) are pairs of port voltage and current vectors related to E'_2, E''_2, E'_1 and E''_1 respectively. P is the matrix that describes the incidence of fundamental cutsets defined by edges in E'_2 and edges in E'_1, Q is the matrix that describes the incidence of fundamental circuits defined by edges in E''_1 and edges in E''_2, and finally, $B_{1''2'}$ is the matrix that describes the incidence of fundamental circuits defined by edges in E''_1 and fundamental cutsets, defined by edges in E'_2. Relations (1') are KVL's written for the fundamental circuits defined by the edges in E'_1. Relations (2'') are KCL's written for the fundamental cutsets defined by the edges in E''_2. The relations (2)' present KCL's for the fundamental cutsets defined by the edges in E'_2. Finally, relations (1)'' present KVL's for the fundamental circuits defined by the edges in E''_1.

We assume that each port of the network is resistive, both v-controlled and i-controlled. Hence, the constitutive relations of network ports will be written either in v-controlled form $i = i(v)$ or in i-controlled form $v = v(i)$:

$$i'_1 = i'_1(v'_1) \tag{b'}$$

$$i'_2 = i'_2(v'_2) \tag{a'}$$

$$v''_1 = v''_1(i''_1) \tag{b''}$$

$$v''_2 = v''_2(i''_2). \tag{a''}$$

The topological relations (1') and (2'') together with the constitutive relations uniquely determine both voltages and currents of each edge of the network in terms of the variables v'_2 and i''_1:

$$v'_2 = v'_2 \tag{$*$}$$

$$i'_2 = i'_2(v'_2) \tag{a'}$$

$$i''_1 = i''_1 \tag{$**$}$$

$$v''_1 = v''_1(i''_1) \tag{b''}$$

$$v'_1 = -P^{\mathrm{T}} v'_2 \tag{1'}$$

$$i'_1 = i'_1(-P^{\mathrm{T}} v'_2) \tag{(b') $\leftarrow$ (1')}$$

$$i_2'' = Q^{\mathrm{T}} i_1'' \tag{2''}$$

$$v_2'' = v_2''(Q^{\mathrm{T}} i_1''). \tag{$(a'') \hookleftarrow (2'')$}$$

Substituting (a′) and ((b′)↩(1′)) into (2′) and substituting (b″) and ((a″)↩(2″)) into (1″), we obtain $|E_2'| + |E_1''|$ relations:

$$i_2'(v_2') = P i_1'(-P^{\mathrm{T}} v_2') + B_{1''2'}^{\mathrm{T}} i_1'' \tag{c'}$$

$$v_1''(i_1'') = -Q v_2''(Q^{\mathrm{T}} i_1'') - B_{1''2'} v_2' \tag{c''}$$

in terms of $|E_2'| + |E_1''|$ topologically independent variables $\{v_2', i_1''\}$. These equations we call the *hybrid network equations* of the network.

Because generally there are several pairs (E_2', E_1'') associated with a partition (E', E''), and there are several choices for the 2-partition (E', E''). We conclude that there are also several choices for the topologically complete set $\{v_2', i_1''\}$ of independent variables with, generally speaking, different cardinalities. The problem of finding a minimum number of independent variables can be reduced to the following problem: find a partition (E', E'') of the edge set E so that the sum $|E_2' + |E_1''|$ is minimal over all partitions (E', E''). Clearly, because $|E_2'| = \operatorname{rank}(N')$ and $|E_1''| = \operatorname{corank}(E'')$, this problem can be preformulated as a problem of finding a partition (E', E'') of the edge set E of a network for which the sum $\operatorname{rank}(E') + \operatorname{corank}(E'')$ is minimal. According to Proposition 3.31, the minimum of the sum $\operatorname{rank}(E') + \operatorname{corank}(E'')$, over all partitions (E', E''), coincides with the hybrid rank of the network. Clearly, E_2' and E_1'' are disjoint subsets of E such that E_2' is circuit-less, E_1'' is cutset-less and $sp_c(E_2') \cup sp_s(E_1'') = E$ (that is, the union of the span of E_2' with respect to circuits and the span of E_1'' with respect to cutsets, cover E). The converse also holds: if (τ, μ) is a pair of disjoint edge subsets of E such that τ is circuit-less, μ is cutset-less and $sp_c(\tau) \cup sp_s(\mu) = E$ then, there is a partition (E', E'') such that in the associated 4-partition $(E_2', E_1', E_2'', E_1'')$, τ coincides with E_2' and μ with E_1''. To see this, note that as such a partition (E', E'') we can take any partition of E satisfying simultaneously the relations: $\tau \subseteq E' \subseteq sp_c(\tau)$ and $\mu \subseteq E'' \subseteq sp_s(\mu)$. Clearly, the voltage of edges in τ and the current of edges in μ, constitute a topologically complete set of variables. Hence, (τ, μ) is a basic pair of subsets. Because $\tau \cap \mu = \emptyset$, in classical hybrid analysis *not more than one* variable (either voltage or current) of an edge can contribute to the topologically complete set of variables. Notice that, using the concept of basic pair, a topologically complete set of variables can be formally defined without referring to a partition (E', E'') of the set E. A basic pair (τ, μ) is said to be *an optimal basic pair* if the sum $|\tau| + |\mu|$ is minimum over all basic pairs of G.

3.9   Procedure for finding an optimal basic pair

In order to construct an efficient (polynomial time) procedure for finding an optimal basic pair (the basic pair which generates a minimal choice of independent variables in hybrid analysis of a network), a version of the general 2-matroid intersection algorithm will be used [60]. In this case one matroid is the graphic matroid and the other is the cographic matroid, both related to the graph G associated with a network.

Before we present the algorithm, we introduce some necessary notation. Given a graph G, denote by P_C the collection of all circuit-less edge subsets of G and by P_S the collection of all cutset-less edge subsets of G. Clearly, an edge subset I of E is a double independent subset of G iff $I \in P_C \cap P_S$. Obviously, the empty set also belongs to $P_C \cap P_S$. Using a procedure similar to that which is described in the proofs of Propositions 3.17 and 3.22, we can augment the empty set in turn, until a maximum cardinality edge subset is obtained. As a tool of this augmenting procedure, we use the so called digraph 'technique'. For a current double independent subset I we construct an auxiliary digraph $D(I)$ for which E serves as the vertex set and $A(I) \subseteq E \times E$ serves as the set of its *arcs*.

Construction of digraph $D(I)$: With each double independent subset I we associate a unique digraph $D(I)$ as follows. By $C(I,x)$ $(S(I,x))$ we denote the unique circuit (cutset) that an edge $x \in E \backslash I$ forms with edges in I only and let $y \in I$. Then $(y,x) \in A$ if $I \cup \{x\} \notin P_C$ and $y \in C(I,x)$. Also, $(x,y) \in A$ if $I \cup \{x\} \notin P_S$ and $y \in S(I,x)$. Thus, edges of G serve as vertices of $D(I)$ and ordered pairs of edges of G serve as arcs of $D(I)$. Because each arc of $D(I)$ connects a member of I with a member of $E \backslash I$, we conclude that $D(I)$ is *directed bipartite* graph.

As an example of constructing the auxiliary digraph $D(I) = (E, A)$ associated with a double independent subset I, consider the graph of Figure 3.11(ii) where members of I are denoted by bold edges. We shall examine in turn all members of the set $E \backslash I = \{a, d, e, f, g\}$. We start with edge a. Because it does not form a circuit with edges in I only, we conclude that it is a target vertex of the auxiliary digraph (belongs to T). At the same time, edge a forms the unique cutset $\{a, b\}$ with edges in I only and hence we add the arc (b, a) to arc set A (which is initially empty). Now we examine edge d. It forms the unique circuit $\{d, c, e\}$ with edges in I only and hence the arcs (d, c) and (d, e) are added to A. Also d forms the unique cutset $\{d, b, c\}$ with edges in I only, so we add arcs (b, d) and (c, d). Therefore, in the auxiliary digraph, d is an ordinary vertex. Edge f does not form a cutset with edges in I only, so it is a source vertex of the auxiliary digraph (belongs to S). On the other hand, f forms a circuit $\{f, e\}$ with edges in I only and hence, (f, e) is added to A. In the same manner, g belongs to S, and (g, e) belongs to A.

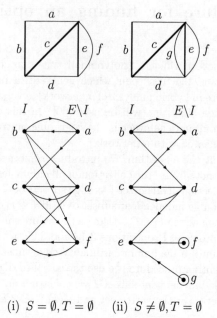

(i) $S = \emptyset, T = \emptyset$ (ii) $S \neq \emptyset, T = \emptyset$

Figure 3.11. Two possible situations of auxiliary digraphs with target vertex sets empty.

If the in-degree (out-degree) of a vertex of $D(I)$ that belongs to $E\backslash I$ is zero, then we call it a *source (target)* vertex. We denote by S (respectively, T) the set of all source (target) vertices in $E\backslash I$ of the auxiliary digraph. If both the in-degree and the out-degree of a vertex in $E\backslash I$ are nonzero, it is an *ordinary* vertex. If both the in-degree and the out-degree of a vertex in $E\backslash I$ are zero, the vertex is a *free* vertex. Thus, a vertex belongs to the intersection $S \cap T$, iff it is a free vertex. If $S \cap T$ is not empty then I can be augmented using any $z \in S \cap T$. Otherwise, if $S \cap T$ is empty, then I is a maximal double independent subset, that is, a basoid. Then an augmenting process can be carried out by looking at the shortest paths of the auxiliary digraph from a vertex in S to a sink vertex in T. As a basis for this process we use the following two corollaries that in 'digraph language' correspond to Propositions 3.17 and 3.22.

Corollary 3.35 *Let $I \in P_C \cap P_S$ and let π be a shortest directed path of the auxiliary digraph $D(I)$ from a vertex in S to a vertex in T and let $W(\pi)$ be its edge set. Then, $I_1 = I \oplus W(\pi) \in P_C \cap P_S$ and $|I_1| > |I|$.*

Corollary 3.36 *I is a maximum cardinality member of $P_C \cap P_S$ (a dyad) iff there is not a directed path of the auxiliary digraph $D(I)$ from a vertex in S to a vertex in T.*

According to Corollary 3.35, if there is a path π of the auxiliary digraph $D(I)$ from a vertex in S to a sink vertex in T we can augment I and obtain a new double independent subset I_1. According to Corollary 3.36 such a path exists iff I is not of maximum cardinality. In each repetition of this procedure we create a new double independent subset with cardinality greater than that of the previous one. Consequently, in turn we obtain a sequence of basoids of strictly increasing cardinalities. Since there are only a finite number of edges in G, the algorithm must terminate with a basoid with maximum cardinality.

Having obtained a dyad (a basoid with maximum cardinality), a procedure for finding its partition which is a basic pair of subsets, depends on whether S and/or T are empty or not. Four typical situations: (i) $S = \emptyset$ and $T = \emptyset$, (ii) $S \neq \emptyset$ and $T = \emptyset$, (iii) $S = \emptyset$ and $T \neq \emptyset$ and (iv) $S \neq \emptyset$ and $T \neq \emptyset$ are presented in Figures 3.11 and 3.12.

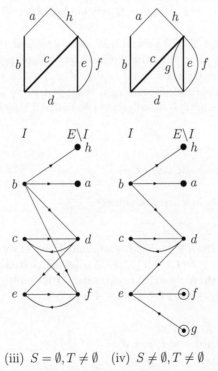

(iii) $S = \emptyset, T \neq \emptyset$ (iv) $S \neq \emptyset, T \neq \emptyset$

Figure 3.12. Two possible situations of auxiliary digraphs with target vertex sets not empty.

Let I be a maximum cardinality member of $P_C \cap P_S$ (that is a dyad) and let $D(I)$ be the associated auxiliary digraph. Let also $S(T)$ be the set of source (target) vertices in $D(I)$ that belong to $E \backslash I$. If M_S is the set of all

members of I that can be reached in the digraph $D(I)$ from source vertices
by simple directed paths then, clearly $sp_c(M_S) \cup sp_s(I \backslash M_S) = E$ (see also
Proposition 3.19). That is, M_S and $I \backslash M_S$ are disjoint subsets of I such that
the span of M_S with respect to circuits and the span of $E \backslash M_S$ with respect
to cutsets, cover E. Hence, the pair $(\tau = M_S, \mu = I \backslash M_S)$ is a basic pair. But
$|M_S| + |I \backslash M_S| = |I|$ and $|I|$ is always less or equal to the hybrid rank of G
(see Lemma 3.30). Therefore the pair $(\tau = M_S, \mu = I \backslash M_S)$ is minimal over
all basic pairs (τ, μ) of G.

Alternatively, if M_T is the set of all vertices in I from which members of T
can be reached in the digraph $D(I)$ by simple directed paths then $sp_c(I \backslash M_T) \cup$
$sp_s(M_T) = E$. Thus, $I \backslash M_T$ and M_T are disjoint subsets of I such that the
span of $I \backslash M_T$ with respect to circuits and the span of M_T with respect to
cutsets, cover E. Hence, the pair $(\tau = I \backslash M_T, \mu = M_T)$ is a basic pair. But
again $|I \backslash M_T| + |M_T| = |I|$ and $|I|$ is less than or equal to the hybrid rank of
G. Hence the pair $(\tau = I \backslash M_T, \mu = M_T)$ is minimal over all basic pairs (τ, μ)
of G.

When $S = \emptyset$ and $T = \emptyset$ we can take as a basic pair of subsets $(\tau = I,$
$\mu = \emptyset)$ or $(\tau = \emptyset, \mu = I)$. Note that besides the basic pairs mentioned above,
there are others that have minimum cardinality.

The following algorithm for finding a basic pair (τ, μ) for which the sum
$|\tau| + |\mu|$ is minimum in G, is based on the two matroid intersection algorithm.

Procedure 2
input: A graph G
output: An optimal basic pair (τ, μ)
Step 1:
$I \leftarrow \emptyset$ (I is a double independent subset)
for all $e \in E \backslash I$ **do**
  begin
    if $I \cup \{e\} \in P_C$ **then** $T \leftarrow T \cup \{e\}$ (initially T is empty)
      else for all $e_1 \in C(I, e) \backslash \{e\}$ $A \leftarrow A \cup \{e_1, e\}$ (initially A is empty)
    if $I \cup \{e\} \in P_S$ **then** $S \leftarrow S \cup \{e\}$ (initially S is empty)
      else for all $e_2 \in S(I, e) \backslash \{e\}$ $A \leftarrow A \cup \{e, e_2\}$.
  end
Step 2:
if $T = \emptyset$ **then output** $\tau = I$, $\mu = \emptyset$
if $S = \emptyset$ **then output** $\tau = \emptyset$, $\mu = I$
if there exists an edge $e \in S \cap T$ **then**
    begin $I \leftarrow I \cup e$, **goto** Step 1 **end**
      else (S and T are nonempty and disjoint)
Step 3:
if there is a directed path between S and T
  then begin

find a shortest directed path π from a member of S
to a member of T, in the auxiliary digraph.
(Let $W(\pi)$ be the subset of E corresponding to π)
$I \leftarrow I \oplus W(\pi)$ (the ring sum of I and $W(\pi)$)
goto Step 1
end
else goto Step 4 (I is the largest member of $P_C \cap P_S$)
Step 4:
In the digraph $D(I)$ find the set M_S of members of I that can be reached
from members of S by simple directed paths.
output $\tau = M_S$, $\mu = I\backslash M_S$.
(Alternatively, find the set M_T of all members of I from which members of T
can be reached in the digraph $D(I)$, by simple directed paths.
output $\tau = I\backslash M_T$, $\mu = M_T$).
end of Procedure 2.

The matroid intersection algorithm correctly finds a maximum subset in
$O(|E|^3 C(|E|))$ time (see, for example, [60]), where $C(|E|)$ is an upper bound
on the complexity of the sub-procedures that decide whether a subset I is
circuit-less or cutset-less. Therefore, the same upper bound holds for Procedure 2.

Clearly, for each graph there is always at least one dyad and, for each dyad
there are generally several different sequences of double independent subsets
that generated it.

As an example consider the graph of Figure 3.1 representing the topology
of a resistive network. Using Procedure 2 (or by inspection) it can be seen
that for this graph ($\tau = \{c, e\}$, $\mu = \{b\}$) is a minimum basic pair. It is easy
to see that $sp_c(\tau) = \{c, d, e, f\}$, $sp_s(\mu) = \{a, b\}$, and hence $sp_c(\tau) \cup sp_s(\mu) =
E$. As a partition (E', E''), which is in accordance with this basic pair, we can
take any partition of E simultaneously satisfying the relations: $\{c, e\} \subseteq E' \subseteq
\{c, d, e, f\}$ and $\{b\} \subseteq E'' \subseteq \{a, b\}$. For example such a partition associated
with ($\tau = \{c, e\}$, $\mu = \{a\}$) is $E' = \{c, d, e, f\}$, $E'' = \{a, b\}$. The voltages
of edges in $\tau = \{c, e\}$ and the current of edges in $\mu = \{a\}$, constitute a
topologically complete set of variables.

3.10   Bibliographic notes

Since the late 1960's many researchers have contributed to important results
concerning maximal subsets that are both circuit-less and cutset-less (either
associated with graphs or with pairs of dual (poly)matroids). However, only
subsets of maximum cardinality received proper attention until recently (Novak and Gibbons [55],[56]). The principal partition with respect to a maximum both circuit-less and cutset-less subset, and many generalizations can be

found through: Ozawa [59], Tomizawa [66], Iri [33] and Chen [10]. The principal partition was formally introduced by Kishi and Kajitani [36] for graphs, in the context of pairs of trees, and then by other authors in more general contexts (see for example, Bruno and Weinberg [4] and Iri and Fujishige [34]). Following Gao and Chen [19], we used the name *dyad* for maximum double independent edge subsets, that is for largest basoids. For the 2-matroid intersection problem see, for example, [37] or [60]. A general version of this algorithm was first described by Lawler [37]. This algorithm finds a basoid of maximum cardinality in polynomial time. It is interesting to note that the dual problem of finding a basoid of minimum cardinality (that is, a minimum cardinality maximal subset independent in two dual matroids) is known to be NP-complete [20]. The notion of *hybrid rank* of a graph was introduced by Tsuchiya *et al.* [67] as the minimum of rank (K) + corank (K^*) over all edge subsets K. In the literature devoted to the analysis of electrical networks, this number is also called the *topological degree of freedom* [67], [57]. We have adopted the term basic pair of subsets from Lin [39]. Various forms of hybrid analysis have been reported in the literature [7], [11], [12] and widely used as topological tools in the formulation of some special type of network equations (see for example [13], [44]). Concerning hybrid topological analysis, Amary [3] first pointed out the possibility of obtaining a minimum number of hybrid equations.

4

Pairs of Trees

Many properties of pairs of trees of a graph are related to the Hamming distance between them. This is important for several graph-theoretical concepts that have featured in hybrid graph theory. Here the notions of perfect pairs and superperfect pairs of trees have played a part. We define and characterize these notions in this chapter and describe necessary conditions for the unique solvability of affine networks in terms of trees and pair of trees.

The small number of theorems and propositions collected together in the opening paragraphs of Chapter 3 will again be frequently referred to here. Familiarity with the basic concepts of graphs such as circuit and cutset are presumed in this chapter. A maximal circuit-less subset of a graph G is called a *tree* of G while a maximal cutset-less subset of edges is called a *cotree*. These terms (circuit, cutset, tree and cotree) will be used here to mean a subset of the edges of a graph. Let F be a subset of E. Then the rank of F, denoted by rank (F), is the cardinality of a maximum circuit-less subset of F and the corank of F, denoted by corank (F), is the cardinality of a maximum cutset-less subset of F. The complement of F is the set difference $E \setminus F$, denoted by F^*. By $|F|$ we denote the number of elements in (that is, the cardinality of) the subset F.

4.1   Diameter of a tree

Given a graph G, each its tree t can be classified according to the non-negative integer rank (t^*). We call this integer the *diameter* of t. Let T denote the set of all trees of a graph G and let $D = \{\text{rank}(t^*)|t \in T\}$ be the set of corresponding diameters. Clearly, D is a finite collection of positive integers. A tree t_c with minimum diameter in G is called a *central tree* and a tree t_e with maximum diameter in G is called an *extremal tree*. Thus, the diameter of t_c is equal to $\min_t D$ and diameter of t_e is equal to $\max_t D$.

An interesting property of the set D is described in the next proposition.

115

Proposition 4.1 *Let G be a graph. Then, the set D of all tree diameters in G is dense in the sense that for any pair of diameters k, j in D such that $k - j \geq 2$, there exists a diameter s in D such that $k > s > j$.*

Before proving Proposition 4.1, we first prove two lemmas.

Lemma 4.2 *Any tree of a graph can be transformed into any other by a sequence of fundamental exchange tree operations.*

Proof Given an arbitrary pair of trees (t_1, t_2) of a graph we define a sequence of pairs of trees (t_1^i, t_2), $i \geq 0$, obtained by consecutive applications of the fundamental exchange tree operation defined as follows. For $i \geq 0$, $t_1^{i+1} = (t_1^i \setminus \{x\}) \cup \{y\}$ where $x \in t_1^i \setminus t_2$, and $y \in t_2 \setminus t_1^i$, where t_1^0 stands for t_1. Clearly $|t_1^{i+1} \setminus t_2| = |t_1^i \setminus t_2| - 1$, that is, after each iteration, the set difference between the current tree t_1^i and t_2 becomes smaller by 1. Eventually, for some $j, |t_1^j \setminus t_2| = 0$ when $t_1^j = t_2$. $\square$

Lemma 4.3 *Let t and t_1 be trees of a graph such that t_1 is obtained from t by applying a fundamental exchange tree operation. Then $(\mathrm{rank}\,(t_1^*) - \mathrm{rank}\,(t^*)) \in \{-1, 0, 1\}$.*

Proof If t_1 is obtained from t by applying a fundamental exchange tree operation, then $t_1 = (t \setminus \{x\}) \cup \{y\}$, where $x \in t \setminus t_1$, $y \in t_1 \setminus t$. Consequently, $t_1^* = (t^* \setminus \{y\}) \cup \{x\}$. Clearly, $\mathrm{rank}\,(t^* \setminus \{y\}) - \mathrm{rank}\,(t^*) \in \{-1, 0\}$ because the substraction of y from t^* cannot decrease the rank by more than 1. Similarly, $\mathrm{rank}\,(t_1^*) - \mathrm{rank}\,(t^* \setminus \{y\}) \in \{0, 1\}$ because the addition of x to $t^* \setminus \{y\}$ cannot increase the rank by more than 1. Hence, $(\mathrm{rank}\,(t_1^*) - \mathrm{rank}\,(t^*)) = [\mathrm{rank}\,(t_1^*) - \mathrm{rank}\,(t^* \setminus \{y\})] + [\mathrm{rank}\,(t^* \setminus \{y\}) - \mathrm{rank}\,(t^*)] \in \{c | c = a + b, \ a \in \{-1, 0\}, b \in \{0, 1\}\} = \{-1, 0, 1\}$.

Proof (of Proposition 4.1) Let t_1 have diameter k and let t_2 have diameter j such that $k - j \geq 2$. Consider the sequence $t_1 = t_1^0, t_1^1, t_1^2, \ldots, t_1^p = t_2$ of trees where any pair of consecutive trees is related by means of a fundamental exchange tree operation. The existence of such a sequence follows from Lemma 4.2. To prove that Proposition 4.1 holds it is enough to note that in the associated sequence of diameters the difference between any two consecutive members is -1, 0 or 1. But this is given by Lemma 4.3. $\square$

For a given graph we can clearly find a tree with any exchange diameter between the diameters of a central tree and an extremal tree.

For example, consider the graph of Figure 4.1. Three copies of this graph with trees made of bold edges and cotrees drawn in dotted edges are shown in Figure 4.2. Bold edges indicate trees and the dotted edges indicate the associated cotrees. The trees have diameters 5, 6 and 7, Figure 4.2(a) shows a central tree and Figure 4.2(c) an extremal tree.

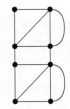

Figure 4.1. A graph.

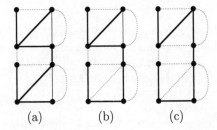

(a)   (b)   (c)

Figure 4.2. Three copies of the graph of Figure 4.1 with trees of diameters 5, 6 and 7, respectively.

Hereafter we present several propositions related to diameters of trees and to the maximal number of cutsets in trees. The next lemma serves as a basis for most of them.

Lemma 4.4 *Given a tree t of a graph, let μ be a maximal cutset-less subset of t and τ be a maximal circuit-less subset of t^*. Then $\tau \cup (t \setminus \mu)$ is a tree. Moreover, it is the unique tree that contains τ as a proper subset and has an empty intersection with μ.*

Proof According to Lemma 2.26, Chapter 2, there is a tree t' such that $\tau \subseteq t'$ and $\mu \cap t'$ is empty. Suppose there is an edge $x \in t \setminus \mu$ that belongs to the complement of t'. Because μ is a maximal cutset-less subset of t, x forms a cutset with edges in μ only, which contradicts the fact that the complement of t' is a cotree. Dually, suppose there is an edge $y \in t^* \setminus \tau$ that belongs to t'. Because τ is a maximal cutset-less subset of t^*, y forms a circuit with edges in τ only, which contradicts the fact that t' is a tree. Hence, $\tau \cup (t \setminus \mu)$ is a tree and there is no other tree containing τ as a proper subset and having an empty intersection with μ.   □

The following proposition is based on Lemma 4.4.

Proposition 4.5 *Let t be a tree of a graph G. Then rank $(t^*) = $ corank (t).*

Proof Let μ be a maximal cutset-less subset of t and let τ be a maximal circuit-less subset of t^*. According to Lemma 4.4, $t' = \tau \cup (t \setminus \mu)$ is also a tree such that $\mu = |t \setminus t'|$ and $\tau = |t' \setminus t|$. But $|t \setminus t'| = |t' \setminus t|$ for any pair of trees and consequently $|\tau| = |\mu|$. But $|\tau| = \text{rank}\,(t^*)$ and $|\mu| = \text{corank}\,(t)$ and hence $\text{rank}\,(t^*) = \text{corank}\,(t)$. $\qquad\qquad\square$

Recall that $n_s(t)$ means the maximal number of independent cutsets of G contained in t.

Proposition 4.6 *Let t be a tree of a graph G. Then $\text{rank}\,(G) = n_s(t) + \text{rank}\,(t^*)$.*

Proof Let μ be a maximal cutset-less subset of t. Then $|\mu| = \text{corank}\,(t)$. On the other hand, any edge in $t \setminus \mu$ forms a cutset with edges in μ only and hence $|t \setminus \mu| = n_s(t)$. Thus $\text{corank}\,(t) + n_s(t) = |t|$. But, from Proposition 4.5, $\text{rank}\,(t^*) = \text{corank}\,(t)$ and thus the relation $\text{rank}\,(G) = n_s(t) + \text{rank}\,(t^*)$ holds. $\qquad\square$

According to Proposition 4.6, $n_s(t)$, associated with a tree t of G is an invariant of the set of all trees with the same diameter. Moreover, the sum $n_s(t) + $ (diameter of t) is an invariant of the graph (that is, an invariant of the set of all trees in the graph). Therefore the diameter of an extremal tree corresponds to the minimum of $n_s(t)$ over all trees in the graph, and the diameter of a central tree corresponds to the maximum of $n_s(t)$ over all trees in the graph. Thus, the following proposition holds.

Proposition 4.7 *Given a graph G, the following statements hold:*

(i) *A tree t_c is a central tree iff $n_s(t_c) = \max_t\{n_s(t)\}$;*

(ii) *A tree t_e is an extremal tree iff $n_s(t_e) = \min_t\{n_s(t)\}$.*

The next two propositions are closely related.

Proposition 4.8 *Given a tree t of a graph G, the following two statements are true:*

(i) $(\forall t') \; |t \setminus t'| \leq \text{rank}\,(t^*)$;

(ii) $(\exists t_0) \; |t \setminus t_0| = \text{rank}\,(t^*)$.

Proof (i) Each subset of a base is independent and its rank is equal to its cardinality. Also, rank is a non-decreasing function. Therefore, the following relations hold for any tree t' of G: $|t' \setminus t| = |t' \cap t^*| = \text{rank}\,(t' \cap t^*) \leq \text{rank}\,(t^*)$.

(ii) Let τ be a maximal circuit-less subset of t^*. Since each circuit-less subset of G can be augmented to a tree, there is a tree t_0, containing τ as a proper subset. Clearly, no edge in $t_0 \setminus \tau$ belongs to t^* because otherwise, we could augment τ with an edge in t^*, which contradicts the assumption that τ is a maximal circuit-less subset of t^*. Thus $t_0 \setminus \tau$ belongs entirely to t and hence, $\tau = t_0 \setminus t$. But $|\tau| = \text{rank}\,(t^*)$ and hence $|t \setminus t_0| = \text{rank}\,(t^*)$. $\qquad\square$

Proposition 4.9 *Let t be a tree of a graph. Then*

(i) *for any tree t' of the graph, $n_s(t) \leq |t \cap t'|$, and*

(ii) *there exists a tree t_0 such that $n_s(t) = |t \cap t_0|$.*

Proof According to Proposition 4.5, rank $(G) = n_s(t) +$ rank (t^*). But rank $(G) = |t| = |t \setminus t'| + |t \cap t'|$, for an arbitrary tree t' in the graph and therefore $|t \setminus t'| -$ rank $(t^*) = n_s(t) - |t \cap t'|$. Hence, $|t \setminus t'| -$ rank $(t^*) \leq 0$ iff $n_s(t) - |t \cap t'| \leq 0$. Also $(\exists t')|t \setminus t'| -$ rank $(t^*) = 0$ iff $(\exists t')n_s(t) - |t \cap t'| = 0$. Thus we have proved that Proposition 4.9 holds iff Proposition 4.8 holds. $\square$

4.2 Perfect pairs of trees

The *distance* between two trees t_1 and t_2 of a graph, written $|t_1 \setminus t_2|$, is defined to be the number of edges which are in t_1 but not in t_2. This type of distance is called the Hamming distance.

The distance between two trees obviously satisfies the axioms of distances, that is:

(i) $|t_1 \setminus t_2| \geq 0$ and $|t_1 \setminus t_2| = 0$ if $t_1 = t_2$.

(ii) $|t_1 \setminus t_2| = |t_2 \setminus t_1|$

(iii) $|t_1 \setminus t_2| + |t_2 \setminus t_3| \geq |t_1 \setminus t_3|$

The first axiom follows from the obvious fact that the cardinality of a finite set is a non-negative integer; it is positive if $t_1 \neq t_2$ and zero if $t_1 = t_2$. The second axiom is obviously true because all trees of a graph have the same cardinality. To prove that the third axiom is true notice the following facts: $t_1 \setminus t_2$ and $t_2 \setminus t_3$ are disjoint subsets and $(t_1 \setminus t_2) \cup (t_2 \setminus t_3) \supseteq (t_1 \setminus t_3)$. These facts immediately imply that $|(t_1 \setminus t_2) \cup (t_2 \setminus t_3)| = |t_1 \setminus t_2| + |t_2 \setminus t_3| \geq |t_1 \setminus t_3|$.

A tree t_2 is said to be *maximally distant from* another tree t_1 if $|t_1 \setminus t_2| \geq |t_1 \setminus t|$ for every tree t of G. The relation 'to be maximally distant from' is generally neither symmetric nor transitive. To see this consider the graph of Figure 4.3 and the three trees t_1, t_2 and t_3 which are indicated with bold edges. It can be seen by inspection that t_2 is maximally distant from t_1 while t_1 is not maximally distant from t_2. Also, t_3 is maximally distant from t_2, t_2 is maximally distant from t_1 but t_3 is not maximally distant from t_1.

The following proposition plays a central role in this section. It gives several equivalent characterizations of the relation 'to be maximally distant from'.

Proposition 4.10 *The following five statements are equivalent:*

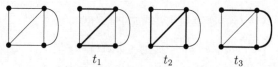

$$t_1 \qquad\qquad t_2 \qquad\qquad t_3$$

Figure 4.3. A graph and three different trees indicated in bold edges.

(i) t_2 *is maximally distant from* t_1.

(ii) *The fundamental circuit with respect to* t_2 *defined by an edge in* $t_1^* \cap t_2^*$ *contains no edges in* $t_1 \cap t_2$.

(iii) *The fundamental cutset with respect to* t_2^* *defined by an edge in* $t_1 \cap t_2$ *contains no edges in* $t_1^* \cap t_2^*$.

(iv) $|t_1 \setminus t_2| = \operatorname{rank}(t_1^*)$.

(v) $n_s(t_1) = |t_1 \cap t_2|$.

Proof (i) $\Rightarrow$ (ii) Suppose that a fundamental circuit with respect to t_2, defined by an edge $c \in t_1^* \cap t_2^*$, contains an edge $b \in t_1 \cap t_2$. Then $t_2' = (t_2 \setminus \{b\}) \cup \{c\}$ is obviously a tree. Because $|t_1 \setminus t_2'| = |t_1 \setminus t_2| + 1$, we conclude that the distance between t_1 and t_2' is larger than that between t_1 and t_2. This contradicts the assumption that t_2 is maximally distant from t_1.

(ii) $\Leftrightarrow$ (iii) This is evident from the following statement (Corollary 2.23, Chapter 2): given a tree t of a graph, an edge e' belongs to a fundamental cutset defined by an edge e'' if and only if e'' belongs to a fundamental circuit defined by the edge e'.

(ii) $\Rightarrow$ (iv) If condition (ii) is satisfied then fundamental circuits with respect to t_2, defined by edges in $t_1^* \cap t_2^*$, contain only edges from $t_2 \setminus t_1$. Hence $\operatorname{rank}(t_1^*) \leq |t_2 \setminus t_1|$. On the other hand, from Proposition 4.8, part (i) $\operatorname{rank}(t_1^*) \geq |t_2 \setminus t_1|$. Therefore $\operatorname{rank}(t_1^*) = |t_2 \setminus t_1| = |t_1 \setminus t_2|$.

(iv) $\Rightarrow$ (i) Suppose that condition (iv) is satisfied, and let t be an arbitrary tree. Then $|t_1 \setminus t_2| = \operatorname{rank}(t_1^*)$ (Condition (iv)) and $|t_1 \setminus t| \leq \operatorname{rank}(t_1^*)$ (Proposition 4.8, part (i)). Consequently, $|t_1 \setminus t| \leq |t_1 \setminus t_2|$.

(iv) $\Leftrightarrow$ (v) According to Proposition 4.6, for any tree t_1 of the graph $\operatorname{rank}(G) = n_s(t_1) + \operatorname{rank}(t_1^*)$. Also for an arbitrary tree t_2 in the graph $\operatorname{rank}(G) = |t_1| = |t_1 \setminus t_2| + |t_1 \cap t_2|$. Therefore $\operatorname{rank}(t_1^*) = |t_1 \setminus t_2| + |t_1 \cap t_2| - n_s(t_1)$ for any pair of trees (t_1, t_2). Consequently, $\operatorname{rank}(t_1^*) = |t_1 \setminus t_2|$ holds iff $|t_1 \cap t_2| = n_s(t_1)$ holds. $\square$

Two copies of the graph of Figure 4.1 with trees indicated in bold edges, are shown in Figure 4.4. The common edges of these two trees and the common edges of the associated cotrees are also marked in the figure. Dotted lines and one curvy solid line indicate cut sets of G belonging to the trees. It can be seen by inspection that there exists a fundamental cutset (shown by

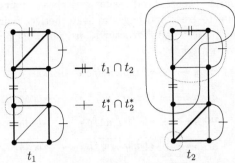

$$\text{-\!\!\!\!-}\quad t_1 \cap t_2$$

$$\text{-\!\!+}\quad t_1^* \cap t_2^*$$

$$t_1 \hspace{5cm} t_2$$

Figure 4.4. A pair of trees illustrating Proposition 4.7.

the curvy solid line) with respect to t_2^*, defined by a common edge of t_1 and t_2 which includes a common edge of the cotrees t_1^* and t_2^*. From Proposition 4.10, part (iii), we conclude that t_2 is not maximally distant from t_1. On the other hand $|t_2 \setminus t_1| = \text{rank}\,(t_2^*)$ and hence (Proposition 4.10, part (iv)) t_1 is maximally distant from t_2.

Propositions 4.8 and 4.10 motivate the name *diameter of a tree*. The diameter of a tree t is the maximum (Hamming) distance between t and any other tree of the graph. There is a difference between this definition of the diameter of a tree and the definition of the diameter of a graph. In the latter case the diameter of a graph is the longest path in the set of shortest paths between pairs of vertices in the graph.

Remark 4.11 If, for a given tree t of a graph, μ is a maximal cutset-less subset of t and τ is a maximal circuit-less subset of t^* then, according to Lemma 4.4, $t' = \tau \cup (t \setminus \mu)$ is also a tree. But $|t' \setminus t| = |\tau| = \text{rank}\,(t^*)$ and consequently (Proposition 4.10), t' is a tree maximally distant from t.

A pair of trees (t', t'') is said to be a *perfect pair of trees* if both t'' is maximally distant from t' and t' is maximally distant from t''.

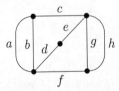

Figure 4.5. A graph having trees of diameters 2, 3 and 4.

In order to provide insight into the structure of perfect pairs, consider the graph in Figure 4.5. This graph has 33 trees in three classes corresponding to diameters 2, 3 and 4. Figure 4.6 describes the perfect pairs structure. Here vertices correspond to trees and edges indicate perfect pairs of trees. Such graphs, describing perfect pairs structures, are generally not connected and may have isolated vertices.

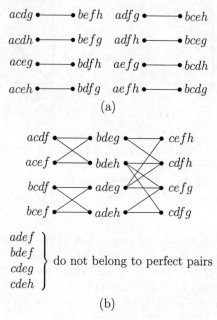

$$
\begin{array}{ll}
acdg \bullet\!\!-\!\!\!-\!\!\bullet befh & adfg \bullet\!\!-\!\!\!-\!\!\bullet bceh \\
acdh \bullet\!\!-\!\!\!-\!\!\bullet befg & adfh \bullet\!\!-\!\!\!-\!\!\bullet bceg \\
aceg \bullet\!\!-\!\!\!-\!\!\bullet bdfh & aefg \bullet\!\!-\!\!\!-\!\!\bullet bcdh \\
aceh \bullet\!\!-\!\!\!-\!\!\bullet bdfg & aefh \bullet\!\!-\!\!\!-\!\!\bullet bcdg
\end{array}
$$

(a)

$$
\begin{array}{lll}
acdf & bdeg & cefh \\
acef & bdeh & cdfh \\
bcdf & adeg & cefg \\
bcef & adeh & cdfg
\end{array}
$$

$$
\left.\begin{array}{l}
adef \\
bdef \\
cdeg \\
cdeh
\end{array}\right\} \text{do not belong to perfect pairs}
$$

(b)

$cdef$ does not belong to a perfect pair

(c)

Figure 4.6. The perfect pairs structure for the graph of Figure 4.5 (a) diameter 4, (b) diameter 3, (c) diameter 2.

As another example consider the graph of Figure 4.7. This has 25 trees in two classes having diameter 2 or 3.

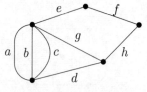

Figure 4.7. A graph.

In order to find a perfect pair of trees of a graph, we can use the following procedure. Let t_1 be an arbitrary tree and let t_2 be a tree maximally distant from t_1 (t_2 may be found by employing Remark 4.11). If t_1 is also maximally distant from t_2, then the pair of trees (t_1, t_2) is a perfect pair. If not, let t_3 be a tree maximally distant from t_2. Then $|t_2 \setminus t_3| > |t_2 \setminus t_1|$ and $|t_1 \setminus t_2| \geq |t_1 \setminus t_3|$. Because $|t_2 \setminus t_1| = |t_1 \setminus t_2|$ we conclude that $|t_2 \setminus t_3| > |t_1 \setminus t_3|$. Hence t_1, cannot be maximally distant from t_3. If t_2 is maximally distant from t_3 then the pair of trees (t_3, t_2) is a perfect pair. If not, let t_4 be maximally distant from t_3. Repeating this procedure we shall obtain a sequence of trees $t_1, t_2, t_3, \ldots$ such that for $k \geq 1$, t_{k+1} is maximally distant from t_k and no tree t_j, where $j < k$, is maximally distant from t_k. Because of the finiteness of a graph, after a number of steps (say p) we obtain t_{p+1} is maximally distant from t_p and vice versa. The pair of trees (t_p, t_{p+1}) is a perfect pair. This procedure defines the algorithm that follows.

Algorithm 1 {To find a perfect pair}
Input (A graph G and a tree t of G)
begin
(1) $t_1 \leftarrow t$
(2) Find a maximal cutset-less subset μ in t and
    find a maximal circuit-less subset τ in t^* ($|\tau| = \text{rank}\,(t^*)$).
(3) $t_2 \leftarrow (t \setminus \mu) \cup \tau$
(4) **if** rank $(t_2^*) = |\tau|$ **then output** (t_1, t_2) is a perfect pair
    else (rank $(t_2^*) > |\tau|$) **begin** $t_1 \leftarrow t_2$, **goto** 1 **end**
end of Algorithm 1

As an immediate consequence of Proposition 4.10, several equivalent characterizations of perfect pairs are possible as the following proposition shows.

Proposition 4.12 *The following five statements are equivalent:*

(i) (t_1, t_2) *is a perfect pair;*

(ii) *Fundamental circuits with respect to t_1 and t_2 defined by edges in $t_1^* \cap t_2^*$ have no edges in $t_1 \cap t_2$;*

(iii) *Fundamental cutsets with respect to t_1^* and t_2^* defined by edges in $t_1 \cap t_2$ have no edges in $t_1^* \cap t_2^*$;*

(iv) rank $(t_1^*) = |t_1 \setminus t_2| = |t_2 \setminus t_1| = \text{rank}\,(t_2^*)$;

(v) $n_s(t_1) = |t_1 \cap t_2| = n_s(t_2)$.

Denote by P the set of all distances between perfect pairs in the graph, and recall the definition of the diameter set D. According to Proposition 4.12, statement (iv), if t_1 and t_2 constitute a perfect pair, then they necessarily have the same diameter which is equal to distance between them. Hence, $D \supseteq P$ and consequently the following relations hold:

(i) $\min P \geq \min D$

(ii) $\max P \leq \max D$.

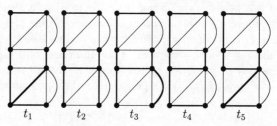

Figure 4.8. Five copies of the graph of Figure 4.1 showing all trees of diameter 4.

Remark 4.13 We shall see later in this chapter (Proposition 4.31), that equality in (ii) always holds. This is not the case with (i), that is, $\min P > \min D$ is possible.

To prove this consider the graph of Figure 4.1. There are exactly five trees with diameter 4 and these are shown in Figure 4.8. It can be checked immediately that the distance between any two of them is equal to 2. Hence, there is no perfect pair with distance 4. In fact, computation shows that the graph of Figure 4.1 has 324 perfect pairs of trees with distance 5, 350 with distance 6 and 63 with distance 7.

The following proposition gives some necessary conditions for a prescribed tree to belong to a perfect pair.

Proposition 4.14 *Let t be a tree of a graph that belongs to a perfect pair of trees. Then necessary conditions for an edge $e \in t$ to be a common edge of the perfect pair as follows:*

(i) *e belongs to a cutset made of edges in t only,*

(ii) *the cutset that e forms with edges in t^* only, is circuit-less.*

Proof Let (t, t_1) be a perfect pair. Then, according to Proposition 4.12, the cutset S_e that an edge in $e \in t \cap t_1$ forms with edges in t_1^* only, contains no edges in $t^* \cap t_1^*$. This means that e forms a cutset with edges in $t \setminus t_1$ only. But $(t \cap t_1) \cup (t \setminus t_1) = t$ and therefore $S_e \subseteq t$. That is, e belongs to a cutset made of edges in t only. On the other hand, the cutset S_e^1 that $e \in t \cap t_1$ forms with edges in t^* only also contains no edges in $t^* \cap t_1^*$. Hence, $S_e^1 \subseteq t_1$ which implies that S_e^1, is circuit-less.   □

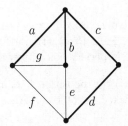

Figure 4.9. A graph illustrating Propositions 4.9 and 4.10.

Remark 4.15 The converse of Proposition 4.14 is generally not true, that is, the assumption that all edges in t satisfy conditions (i) and (ii) is not sufficient for the existence of a perfect pair that includes t.

To see this consider the graph of Figure 4.9 with the tree t indicated with bold edges. It can be seen that each edge in t satisfies conditions (i) and (ii). Nevertheless there is no perfect pair that includes t.

We now prove a much stronger result than Proposition 4.14.

Proposition 4.16 *Let t be a tree of a graph and let π be a subset of t. Then π is a set of all the common edges of a perfect pair of trees that includes t iff the following conditions are fulfilled:*

(i) *$t \setminus \pi$ is a maximal cutset-less subset of t,*

(ii) *the union of all the cutsets, that edges in π form with edges in t^* only, is circuit-less.*

Proof Let (t, t_1) be a perfect pair of trees and let $\pi = t \cap t_1$. Then $t \setminus \pi = t \setminus t_1$ is a proper subset of t_1^* and, consequently, is cutset-less. To prove that it is a maximal cutset-less subset of t it is sufficient to show that any edge in π forms a cutset with edges in $t \setminus \pi$ only, which obviously follows from Proposition 4.12. Using similar reasoning, we conclude that every edge in π forms a cutset with edges in $t_1 \setminus t$ only. Because $(t_1 \cap t) \cup (t_1 \setminus t) \equiv t_1$, any of these cutsets belongs entirely to t_1 and hence is circuit-less. But the union of subsets of a tree still belongs to that tree and so the union of cutsets that edges in π form with edges in $t_1 \setminus t$ is circuit-less.

Conversely, suppose that conditions (i) and (ii) are satisfied for a subset π of t. Let ω be the union of all cutsets that edges in π form with edges in t^* only. According to (ii), $t^* \cap \omega$ is circuit-less. Clearly, $t^* \cap \omega$ can be extended to a maximal circuit-less subset of t^*. Denote by τ this maximal circuit-less subset of t^* that includes $t^* \cap \omega$. Then, according to Lemma 4.4, $t_1 = \tau \cup \pi$ is a tree. By construction, the cutsets that an edge in π forms with edges in t^* entirely belongs to t_1, that is, contains no edges in $t^* \cap t_1^*$ (*fact 1*). On the other hand, according to condition (i), $t \setminus \pi$ is a maximal cutset-less subset of t and hence every edge in π forms a cutset with edges in $t \setminus \pi$ only. But $t \setminus \pi$

and therefore the cutset that an edge in π makes with edges in t_1^* contains no edges in $t^* \cap t_1^*$ (*fact 2*). From Proposition 4.10 the conjunction of *facts 1 and 2* ensures that (t, t_1) is a perfect pair of trees. □

Remark 4.17 Condition (i) of Proposition 4.16 implies condition (i) of Proposition 4.14 while the converse is not true. Also, condition (ii) of Proposition 4.16 is much stronger than condition (ii) of Proposition 4.14 because circuit-lessness of the union of subsets of a graph implies the circuit-lessness of each subset in the union, but the converse is not true. For example, subsets $\{a, f, g\}$ and $\{d, e, f\}$ of the graph of Figure 4.9 are circuit-less but their union is not.

Remark 4.18 Consider again the graph of Figure 4.9. Obviously there is no subset of t that satisfies both conditions (i) and (ii) of Proposition 4.14. Thus any of the edges a, b, c or d alone satisfies condition (ii), but not condition (i). Any pair of these edges except $\{a, b\}$ satisfies condition (i), but not condition (ii). The pair $\{a, b\}$ and any triple or quadruple chosen from $\{a, b, c, d\}$ do not satisfy either (i) or (ii).

4.3   Basoids and perfect pairs of trees

Recall the notion of double independent edge subset and that of basoid. The following lemma connects double independent edge subsets with pairs of trees:

Lemma 4.19 *A subset d of the edge set E of a graph is a double independent subset iff it can be represented as a set difference of a pair of trees of the graph.*

Proof Let d be a double independent subset of a graph G and let t_1 be a tree that contains d as a proper subset (this is always possible since d is circuit-less). Let t_2 be a maximal circuit-less subset of d^* that contains $t_1 \setminus d$ as a proper subset Because d is also cutset-less subset of G, it belongs to a cotree of G and hence rank $(d^*) = $ rank (E). Consequently, every maximal circuit-less subset of d^* is a tree of G, including t_2. But t_2 contains $t_1 \setminus d$ as a proper subset and hence $t_1 \setminus t_2 = d$.

Conversely, let (t_1, t_2) be a pair of trees of a graph G. Then, $t_1 \setminus t_2$ is a subset of both t_1 and t_2^*. Therefore, $t_1 \setminus t_2$ is a double independent subset of G. □

For example, Figure 4.10 shows four copies of the same graph with a pair of trees and their set differences, indicated using bold lines. By inspection, we see that both $t_1 \setminus t_2$ and $t_2 \setminus t_1$ are double independent subsets of the graph.

The following lemma gives necessary and sufficient condition for a set difference of a pair of trees to be a maximal double independent edge subset (basoid).

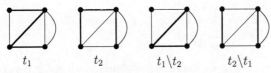

t_1 $\qquad$ t_2 $\qquad$ $t_1 \backslash t_2$ $\qquad$ $t_2 \backslash t_1$

Figure 4.10. Four copies of the graph of Figure 4.3 with a pair of trees and their set differences.

Lemma 4.20 *A set difference $t_2 \setminus t_1$ of a pair of trees (t_1, t_2) is a basoid iff any fundamental cutset that edges in $t_1 \cap t_2$ form with edges in t_2^* and any fundamental circuit that edges in $t_1^* \cap t_2^*$ form with edges in t_1 are disjoint.*

Proof Given a pair of trees (t_1, t_2) of a graph G, let $b = t_2 \setminus t_1$ be a basoid of G. Clearly, every fundamental cutset that an edge in $t_1 \cap t_2$ forms with edges in t_2^* is a cutset made of edges in $b^* = (t_2 \setminus t_1)^*$ only. Also every circuit that an edge in $t_1^* \cap t_2^*$ forms with edges in t_1 is a circuit made of edges in $b^* = (t_2 \setminus t_1)^*$ only. Since b is a basoid, according to Proposition 3.6, part (ii) (Chapter 3), any circuit made of edges in b^* only and any cutset made of edges in b^* only are disjoint. Hence, any fundamental cutset that edges in $t_1 \cap t_2$ form with edges in t_2^* and any circuit that edges in $t_1^* \cap t_2^*$ form with edges in t_1 are disjoint.

Conversely, given a pair of trees (t_1, t_2) of a graph G, suppose that any fundamental cutset that edges in $t_1 \cap t_2$ form with edges in t_2^* and any circuit that edges in $t_1^* \cap t_2^*$ form with edges in t_1 are disjoint. Let C be a circuit made of edges in $(t_2 \setminus t_1)^*$. Certainly, C must contain at least one edge in $t_1^* \cap t_2^*$ because otherwise $C \subseteq t_1$ (which is a contradiction). Suppose C contains an edge $x \in t_1 \cap t_2$. Because x belongs to t_1, according to Corollary 2.29, Chapter 2, there exists $y \in C \cap (t_1^* \cap t_2^*)$ such that the fundamental circuit C_y defined by y with respect to t_1 contains x. But x forms a cutset S_x with edges in t_2^* only and hence x belongs to both C_y and S_x, which contradicts our supposition. Thus we have proved that any circuit in $(t_2 \setminus t_1)^*$ is a proper subset of t_2^*, having at least one edge in $t_1^* \cap t_2^*$. In a dual manner, we can prove that any cutset in $(t_2 \setminus t_1)^*$ is a proper subset of t_1 having at least one edge in $t_1 \cap t_2$. Suppose now that there are a circuit C in t_2^* and a cutset S in t_1 such that their intersection $C \cap S$ is not empty. Let z be an edge that belongs to $C \cap S$. Then, according to Corollary 2.29, Chapter 2 and its dual, there is an edge $y \in C \cap (t_1^* \cap t_2^*)$ that forms a fundamental circuit with edges in t_1, including z, and there is an edge $x \in S \cap (t_1 \cap t_2)$ that forms a fundamental cutset with edges in t_2^*, including z. This is a contradiction which completes the proof. $\square$

The next proposition provides a link between basoids and perfect pairs of trees.

Proposition 4.21 *Let b be a basoid of a graph. Then there exists a perfect pair (t_1, t_2) such that $b = t_1 \setminus t_2$.*

Proof Because b is by definition a double independent subset of the graph, according to Lemma 4.19, there exists a pair of trees (t_1, t_2) such that $b = t_1 \setminus t_2$. We now show that (t_1, t_2) is a perfect pair. From Proposition 3.6, part (iii) of Chapter 3, each edge of the complement of this basoid (including the edges in $t_1^* \cap t_2^*$) makes a circuit or/and a cutset with the elements of $t_1 \setminus t_2$ only. But $t_1 \setminus t_2$ together with $t_1^* \cap t_2^*$ belongs to t_2^* and hence the edges in $t_1^* \cap t_2^*$ cannot make cutsets with the edges in $t_1 \setminus t_2$ only. Therefore the edges in $t_1^* \cap t_2^*$ make circuits with edges in $t_1 \setminus t_2$ only and consequently rank $(t_2^*) = |t_1 \setminus t_2|$. This means that t_1 is maximally distant from t_2. On the other hand, from Proposition 4.8, part (i), rank $(t_1^*) \geq |t_1 \setminus t_2|$. We now prove that equality occurs in the present case, that is, t_2 is maximally distant from t_1. Suppose that this is not true. Then from Proposition 4.10, part (ii), there exists an edge $e' \in t_1^* \cap t_2^*$ such that a fundamental circuit with respect to t_2 defined by e' contains an edge $e \in t_1 \cap t_2$. Then $t_2' = (t_2 \setminus e) \cup \{e'\}$ is also a tree and $t_1 \setminus t_2 \subseteq t_1 \setminus t_2'$. But, according to Lemma 4.19, the subset $t_1 \setminus t_2'$ is also a double independent subset, that contains as a proper subset the basoid $t_1 \setminus t_2$. Thus we have a contradiction and so t_2 is maximally distant from t_1. Hence (t_1, t_2) is a perfect pair.   □

Remark 4.22 The converse of Proposition 4.21 is generally not true. That is, if (t_1, t_2) is a perfect pair of trees then their set difference is not necessarily a basoid.

To see this consider Figures 4.10 and 4.11. Figure 4.10 shows four copies of the same graph and within each a subset of edges is indicated using bold lines. Now (t_1, t_2) is a perfect pair and (by inspection) $t_1 \setminus t_2$ is a basoid, while $t_2 \setminus t_1$ is not. Figure 4.11 shows four copies of the same graph and again various subsets of edges are indicated using bold lines. Again (t_1, t_2) is a perfect pair, while neither $t_1 \setminus t_2$ nor $t_2 \setminus t_1$ is a basoid. The marked edges (indicating $t_1 \setminus t_2$ and $t_2 \setminus t_1$) form neither circuits nor cutsets.

Proposition 4.23 *Given a perfect pair of trees (t_1, t_2) of a graph G, let $b = t_1 \setminus t_2$ be a basoid. Then $t_2 \setminus t_1$ is a maximally double independent subset in b^*.*

Proof The subset $(t_2 \setminus t_1)$ is obviously double independent (Lemma 4.19) and belongs to b^*. Because (t_1, t_2) is a perfect pair of trees, according to Proposition 4.12, every edge in $t_1^* \cap t_2^*$ forms a circuit with edges in $t_2 \setminus t_1$ only and every edge in $t_1 \cap t_2$ forms a cutset with edges in $t_2 \setminus t_1$ only. Hence, every edge in b^* that does not belong to $t_2 \setminus t_1$ forms a circuit or a cutset with

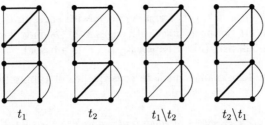

$$t_1 \qquad\qquad t_2 \qquad\qquad t_1\backslash t_2 \qquad\qquad t_2\backslash t_1$$

Figure 4.11. Four copies of the graph of Figure 4.1 with a pair of trees and their set differences indicated in bold edges.

edges in $t_2 \backslash t_1$ only. Consequently, $t_2 \backslash t_1$ is a maximally double independent subset in b^*.   □

The next proposition gives necessary and sufficient conditions for the set difference of a perfect pair of trees to be a basoid.

Proposition 4.24 *Let (t_1, t_2) be a perfect pair of trees of a graph. Then $t_1 \backslash t_2$ is a basoid iff each edge in $t_2 \backslash t_1$ that belongs to the fundamental circuit defined by an edge in $t_1^* \cap t_2^*$ (with respect to t_2), forms a circuit with edges in $t_1 \backslash t_2$ only.*

Proof Suppose $t_1 \backslash t_2$ is a basoid and let x be an edge in $t_2 \backslash t_1$ that belongs to the fundamental circuit $C(y, t_2)$ defined by an edge y in $t_1^* \cap t_2^*$ (with respect to t_2). Because $t_1 \backslash t_2$ is a basoid, x forms a circuit or a cutset with edges in $t_1 \backslash t_2$ only. But x belongs to $C(y, t_2)$, which is made of edges in $(t_1 \backslash t_2)^*$ only, and consequently x necessarily forms a circuit with edges in $t_1 \backslash t_2$ only (otherwise, if x forms only a cutset S with edges in $t_1 \backslash t_2$ only then $C(y, t_2) \cap S = \{x\}$, which, according to the Orthogonality Theorem, is a contradiction).

Conversely, suppose that each edge in $t_2 \backslash t_1$ that belongs to a fundamental circuit with respect to t_2, defined by an edge in $t_1^* \cap t_2^*$, forms a circuit with edges in $t_1 \backslash t_2$ only. Let us denote by R the set of edges in $t_2 \backslash t_1$, that do not belong to circuits that edges in $t_1^* \cap t_2^*$ make with edges in t_2 only. We prove that the edges in R do not belong to any circuit of edges in the complement of $t_1 \backslash t_2$. The proof is by contradiction. Suppose there is a circuit C made of edges in $(t_1 \backslash t_2)^*$ containing an edge x which belongs to R. Certainly, C must contain at least one edge in $t_1^* \cap t_2^*$ because otherwise $C \subseteq t_2$. According to Corollary 2.29, Chapter 2, there exists $y \in C \cap (t_1^* \cap t_2^*)$ such that the fundamental circuit C_y defined by y with respect to t_2 contains x, which contradicts our supposition that x belongs to R. By applying the Painting Theorem to the triple $((t_1 \backslash t_2)^* \backslash \{y\}, \{y\}, t_1 \backslash t_2)$, we conclude that every edge in R necessarily forms cutsets with edges in $t_1 \backslash t_2$ only. Thus every edge in $t_2 \backslash t_1$ forms either a circuit or a cutset with edges in $t_1 \backslash t_2$ only. On the other hand, since the pair (t_1, t_2) is a perfect pair, from Proposition 4.12, part (ii) and (iii), every edge in $t_1^* \cap t_2^*$ forms a circuit with edges in $t_1 \setminus t_2$ only and every edge in $t_1 \cap t_2$ forms a cutset with edges in $t_1 \setminus t_2$ only. Thus, according to Proposition 3.6, part (iii) (Chapter 3), $t_1 \setminus t_2$ is a basoid.   □

Proposition 4.25 is an immediate consequence of Propositions 4.21 and 4.24.

Proposition 4.25 *A subset of edges b of a graph is a basoid iff the conjunction of the following two statements hold:*

(i) *there exists a perfect pair (t_1, t_2) of the graph such that $b = t_1 \setminus t_2$,*

(ii) *each edge in $t_2 \setminus t_1$ that belongs to a fundamental circuit with respect to t_2, defined by an edge in $t_1^* \cap t_2^*$, forms a circuit with edges in $t_1 \setminus t_2$ only.*

Remark 4.26 Conditions (i) and (ii) are mutually independent. In other words neither (i) $\Rightarrow$ (ii) nor (ii) $\Rightarrow$ (i).

To see that (i) does not imply (ii), it is sufficient to consider the example of Figure 4.11. Although the pair (t_1, t_2) is a perfect pair, condition (ii) is not satisfied. Conversely, in order to see that (ii) does not imply (i), consider the example of Figure 4.12. By inspection, condition (ii) is satisfied but rank $(t_1^*) = 5 > 4 = |t_1 \setminus t_2|$ and hence (t_1, t_2) is not a perfect pair.

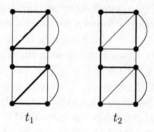

$$t_1 \qquad\qquad\qquad t_2$$

Figure 4.12. Two copies of the graph of Figure 4.1 with a pair of trees indicated in bold edges.

4.4   Superperfect pairs of trees

A pair of trees (t_1, t_2) is defined to be a *superperfect pair of trees* if any fundamental cutset that an edge in $t_1 \cap t_2$ forms with edges in t_1^* or t_2^* and any fundamental circuit that an edge in $t_1^* \cap t_2^*$ forms with edges in t_1 or t_2, are disjoint.

As an example, consider again the graph of Figure 4.1. Here all superperfect pairs can be placed into one of two classes according to their distance apart: those with distance 6 and those with distance 7. This is one of the smallest examples with at least two different distances. Representatives, one for each of these classes, are given in Figure 4.13. By computation, the graph of Figure 4.1 has 32 superperfect pairs of trees with distance 6 and 63 with distance 7.

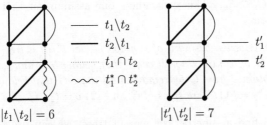

Figure 4.13. Two copies of the graph of Figure 4.1 with superperfect pair of trees with distances 6 and 7.

Starting with the graph of the above example, it is possible to build more complicated graphs by stringing together copies of a basic unit. Figure 4.1 is a string of two such units whereas the example of Figure 4.14 is a string of n units. It is not difficult to prove that all superperfect pairs of the graph of n units can be classified by cardinality into n subsets with distances $3n, 3n+1,$ $\ldots, 4n-1$.

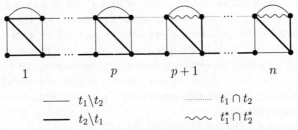

Figure 4.14. A template which allows construction of graphs with superperfect pairs of trees with a prescribed range of distances.

The next proposition, which is an immediate consequence of Lemma 4.20, provides a link between basoids and superperfect pairs of trees.

Proposition 4.27 *A pair of trees (t_1, t_2) is a superperfect pair iff both set differences $t_1 \setminus t_2$ and $t_2 \setminus t_1$ are basoids.*

The following proposition provides a link between perfect pairs of trees and superperfect pairs of trees.

Proposition 4.28 *A superperfect pair of trees is a perfect pair.*

Proof Given a superperfect pair of trees (t_1, t_2), suppose that (t_1, t_2) is not a perfect pair. That is, suppose that there is an edge x in $t_1^* \cap t_2^*$ such that the fundamental circuit C_x that x forms with edges in t_1 or t_2 contains an edge y in $t_1 \cap t_2$. Without loss of generality assume that C_x is the fundamental circuit with respect to t_2. Since y forms the cutset S_y with edges in $t_1^* \subseteq (t_1 \setminus t_2)^*$, x belongs to $C_x \cap S_y$. This contradicts our assumption that (t_1, t_2) is a superperfect pair. $\square$

Remark 4.29 *Given a basoid b of a graph G, let t'' be a tree of the graph obtained from G by contracting all edges that belong to b and let t' be a tree of the graph obtained from G by removing all edges G that belong to b. Then (1) $t_1 = t'$ and $t_2 = b \cup t''$ are trees of G and (2) $b = t_2 \setminus t_1$.*

In order to find a superperfect pair of trees, we can use the following procedure, starting with an arbitrary basoid b_0 of the graph. According to Remark 4.29 we can construct a pair of trees (t_1, t_2) such that $b_0 = t_1 \setminus t_2$. If $t_2 \setminus t_1$ is also a basoid then, according to Proposition 4.27, (t_1, t_2) is a superperfect pair of trees and the algorithm terminates. On the other hand if $t_2 \setminus t_1$ is not a basoid, according to Lemma 4.19, it is still double independent. In this case we add edges to $t_2 \setminus t_1$ until we obtain a basoid, b_1 say, in the manner described earlier. Again, according to Remark 4.29, we construct a pair of trees (t_1', t_2') such that $b_1 = t_1' \setminus t_2'$. If $t_2' \setminus t_1'$ is also a basoid then, again according Proposition 4.27, (t_1', t_2') is a superperfect pair. If not then we continue in repetitive manner until we find a pair of trees for which both set differences are basoids. Termination is guaranteed after a finite number of repetitions by the following observations. In each repetition we create a new pair of trees with a distance greater than that of the previous one. Consequently, in turn we obtain a sequence of basoids of strictly increasing cardinality $|b_0| < |b_1| < |b_2| \dots$. There are only a finite number of edges in G and so the algorithm must terminate with a pair of trees for which both set differences are basoids.

Thus we obtain the following algorithm:

Algorithm 2 (to find a superperfect pair of a graph)
Input: (a graph G and a pair of trees (t_1, t_2) of G)
begin
1    $b_0 \leftarrow \emptyset$
     $f \leftarrow$ set of all edges of the graph G
     $b \leftarrow b_0$
     while f is nonempty **do**
              begin
              Contract the edges in G that belong to b and in the graph
              obtained identify edges that belong to self circuits
              Denote the set of these edges by C

> Remove the edges in G that belong to b and in the graph
> obtained identify edges that belong to self cutsets. Denote
> the set of these edges by S
> $f \leftarrow (b \cup C \cup S)^*$
> Choose any edge $e \in f$
> $b \leftarrow b \cup \{e\}$
> **end**

2 Contract all edges in G that belong to b and find a tree t'' of
   the graph obtained. Remove all edges in G that belong to b and
   find a tree t' of the graph obtained
   $t_2' \leftarrow b \cup t''$
   $t_1' \leftarrow t'$ {**comment**: $b = t_2' \backslash t_1'$}

3 Contract all edges in G that belong to $t_1' \backslash t_2'$. Let C' be the set
   of all self-circuits. Remove all edges in G that belong to $t_1' \backslash t_2'$
   Let S' be the set of all self-cutsets
   If $C' \cup S' \cup (t_1' \backslash t_2')$ covers the edge set of G
   > **then output** $t_1' \backslash t_2'$ is a basoid and (t_1', t_2') is a superperfect
   > pair of trees
   > **else begin** $b_0 \leftarrow t_1' \backslash t_2'$, **goto** 1 **end**

end of Algorithm 2

The next remark clarifies some aspects of Algorithm 2.

Remark 4.30 Let (t_1, t_2) be a perfect pair of trees of a graph for which $t_1 \backslash t_2$ is not a basoid. Let (t_1', t_2') be another perfect pair such that $t_1' \backslash t_2'$ is a basoid containing $t_1 \backslash t_2$ as a proper subset. According to Proposition 4.12, we cannot enlarge $t_1 \backslash t_2$ with elements of $t_1^* \cap t_2^*$ or $t_1 \cap t_2$. So, we have to take some edges from $t_2 \backslash t_1$. This means that at least one edge from $t_2 \backslash t_1$ belongs to $t_1' \backslash t_2'$. But $t_1' \backslash t_2'$ and $t_2' \backslash t_1'$ are disjoint and consequently $t_2 \backslash t_1$ only partly belongs to $t_2' \backslash t_1'$. As a consequence we conclude that $t_2 \backslash t_1$ does not belong to $t_2' \backslash t_1'$ as a proper subset.

Let D denote the set of all tree diameters of a graph G, P the set of all perfect pair distances of G and S the set of all superperfect pair distances of G. From Proposition 4.12, part (iv), the diameter of t_1 is equal to the distance between t_1 and t_2 which is equal to the diameter of t_2. Hence, P is a subset of D. According to Proposition 4.28, S is a subset of P. Thus, $S \subseteq P \subseteq D$. It follows immediately that $\min S \geq \min P \geq \min D$ and $\max S \leq \max P \leq \max D$.

We have already shown (see Remark 4.13) that the strong inequality $\min P > \min D$ can occur. We shall see that the strong inequality $\min S > \min P$ can also occur. To prove this consider the graph of Figure 4.3. As shown in Example 3.1 of Chapter 3, for this graph there is only one subset of cardinality 2 which is basoid. According to Proposition 4.27 this implies that there is not a superperfect pair of trees of this graph of distance 2. On the other hand it can be seen by inspection that (t_1, t_2) is a perfect pair of trees of distance 2.

The next proposition shows that $\max S$, $\max P$ and $\max D$ coincide.

Proposition 4.31 *For every graph* G, $\max S = \max P = \max D$.

Proof The proof is by construction. Suppose that $\max S < \max D$ for a graph G and let (t_1, t_2) be a pair of trees with maximum distance in the graph. According to our assumption (t_1, t_2) is not a superperfect pair of trees and therefore, from Proposition 4.27, at least one of the set differences $t_1 \setminus t_2$ and $t_2 \setminus t_1$ is not a basoid. Suppose, this is the case with the set difference $t_2 \setminus t_1$. Then following Algorithm 2 we can always find a superperfect pair of trees with a distance greater than that of the original pair of trees. This contradicts the assumption that (t_1, t_2) is a pair of trees with maximum distance in G. Consequently we have proved that $\max S = \max D$. This relation together with the relation $\max S \leq \max P \leq \max D$ implies that $\max S = \max P = \max D$. $\qquad\square$

4.5 Unique solvability of affine networks

For many problems, especially those concerned with lumped circuits, trees and pairs of trees of a graph play an important role. In this section we present some graph-oriented criteria for testing the necessary conditions for the unique solvability of so-called time-invariant affine, resistive networks. Each of these criteria is described in terms of a tree or a pair of trees. We define affine networks below and what we mean by their unique solvability.

Let F be a field and let $(V, + \mid \mathsf{F})$ be a linear (vector) space over F. Following the usual notation we define $H + K = \{h + k \mid h \in H \text{ and } k \in K\}$, for any two arbitrary nonempty $H, K \subset V$. The fact that a subset A_o of V is a linear subspace of the linear space V is denoted by $A_o \subset\subset V$. If for $B \subset V$ the relation $(B - B) \subset\subset V$ holds we say that B is an *affine space* (affine manifold) over F. As is well known, for each affine subset B of V there is a unique $B_o \subset\subset V$ such that for every $b \in B$, $B - \{b\} = B - B = B_o$. We also say that B is parallel to B_o. A pair (A_o, B) of subsets of a vector space V is said to be an *abstract network* over the vector space V if A_o is a linear and B is an arbitrary subset of V. An abstract network is *affine* if B is affine. The intersection $A_o \cap B$ represents the *set of solutions* of the abstract network. If this intersection $A_o \cap B$ is nonempty we say that the abstract network is *noncontradictory* (it has at least one solution). If $|A_o \cap B| = 1$, we say that the abstract network has a *unique solution*. It can be shown, that a noncontradictory abstract affine network (A_o, B) has a unique solution iff $|A_o \cap (B - B)| = 1$, and $B \subset A_o + (B - B)$. Notice that condition $|A_o \cap (B - B)| = 1$ alone is not sufficient to guarantee the existence of a solution of (A_o, B). To see this consider the affine abstract network (A_o, B) associated with the network of Figure 4.15. If $E \neq 0$, the network is contradictory because of the presence of the nullator, and therefore (A_o, B) is also contradictory although $|A_o \cap (B - B)| = 1$ holds.

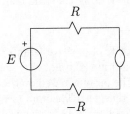

Figure 4.15. A resistive network illustrating some aspect of unique solvability.

As we mentioned in Section 1.8, every network is closed by means of couplings, and hence its collection of ports can be always treated as one global multiport. We describe its constitutive relation by the set B of all $2b$-tuples of signals $(u_1(t), i_1(t); \ldots; u_b(t), i_b(t))$ that are allowed on its ports. This set includes all information concerning individual ports as well as their couplings. Similarly, galvanic connections can be described by a set A_o of $2b$-tuples of signals $(u_1(t), i_1(t); \ldots; u_b(t), i_b(t))$ that are allowed by Kirchhoff's laws. As is known, A_o is linear over F. Thus, for each network, we have two sets of allowed $2b$-tuples of signals: one describing the topology of the network (set A_o linear over F) and the other describing individual ports of the network and their possible couplings (the set B). The pair (A_o, B) is an abstract model of the network. In the case of resistive networks, both A_o and B can be treated as subsets of F^{2b}, where A_o is a linear and B is any subset of F^{2b}. Notice that F^{2b} itself can be treated as a linear space over F.

Consider networks consisting of Ohm's resistors and independent voltage and current sources only. Short circuits will be treated as independent voltage sources and open circuits as independent current sources. For such networks the following well-known condition is necessary for unique solvability:

(i) *there is a tree containing all independent voltage sources and none of the independent current sources.*

The converse is generally not true. Consider the network of Figure 4.16 in which all parameters are nonzero. When $R' + R'' \neq 0$, the network has a unique solution. But if $R' + R'' = 0$ the network is contradictory, although condition (i) holds. Because both R' and R'' are nonzero, $R' + R'' = 0$ implies that either R' or R'' has a negative value.

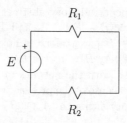

Figure 4.16. A network illustrating that the converse of (i) generally is not true.

For a network that consists of positive Ohm's resistors and constant voltage and current sources, condition (i) is not only necessary but is also sufficient for its unique solvability. The proof is based on the Painting Theorem for partially directed graphs and Tellegen's Theorem (see Assertion 1.39 and 1.42 of Chapter 1; the proof of Corollary 1.42 is given in Chapter 2 as Proposition 2.41).

Proposition 4.32 *For a network made of positive Ohm's resistors and constant independent sources, condition* (i) *is sufficient for unique solvability.*

Proof Consider a network that consists of b ports, $k = 1, \ldots, b$, each of which is either a positive Ohm's resistor ($u_k = R_k i_k, R_k > 0$ or $i_k = G_k u_k$, $G_k > 0$), or a constant independent source ($u_k = E_k$ or $i_k = I_k$) where (u_k, i_k) is the voltage-current pair associated with the port k. Because the network is regular (the number of ports is equal to the number of constitutive relations), to prove that the network has a unique solution it is sufficient to prove that the network obtained by replacing every voltage source E_k by a short circuit and replacing every current source I_k by an open circuit has a unique solution. Thus each port k of the modified network is either a positive Ohm's resistor ($u_k i_k = R_k i_k^2 \geq 0$) or a short circuit ($u_k i_k = 0$) or an open circuit ($u_k i_k = 0$). According to Tellegen's Theorem, applied to this network, the following relation holds: $\sum_{k=1}^{b} u_k i_k = 0$.

Since $u_k i_k \geq 0$, we conclude that $u_k i_k = 0$ for all k. That is, for every k at least one of the port variables u_k and i_k is zero. In the case of Ohm's resistors this implies that both port variables u_k and i_k are equal to zero. Since the voltage of every short circuit is always zero and since the current of every open circuit is always zero, to prove that the modified network has a unique solution it is sufficient to prove that the current of every short circuit and the voltage of every open circuit are also zero. Suppose that there are ports for which the currents are not zero. Clearly, in this case, they must be short circuit ports. Divide the network ports into three classes: P_{io} (short circuit ports for which the currents are not equal to zero), P_i' (short circuit ports for which the currents are equal to zero) and P_i'' (the remaining ports, which are Ohm's resistors and open circuits). Note that $P_{io} \cup P_i'$ contains only short circuit ports and that $P_i' \cup P_i''$ contains all ports for which the current is equal to zero. Suppose that every port k in P_{io} is oriented such that the $i_k > 0$. Let G be a partially oriented graph associated with the network in which the edges that correspond to ports in P_{io} are oriented in accordance with the actual current orientations and the remaining edges are not oriented. According to the Painting Theorem for partially directed graphs, each edge in P_{io} either belongs to a uniform circuit made of edges in $P_{io} \cup P_i'$ only or to a uniform cutset made of edges in $P_{io} \cup P_i''$ only. But every circuit made of edges in $P_{io} \cup P_i'$ only is a circuit made of short circuit ports. This contradicts condition (i). Hence, every edge k in P_{io} belongs to a uniform cutset made of edges in $P_{io} \cup P_i''$ only. Since the cutset is uniform, in the associated Kirchhoff's current law all members appear with a positive sign. On the other hand every member of the sum is either positive (when it is related to the ports in P_{io}) or zero. According to our assumption P_{io} is nonempty and hence at least one member of the sum is nonzero. This is a contradiction. Thus we conclude that P_{io} is empty, that is all short circuit ports belong to P_i'. Suppose now that there are ports for which the voltages are not zero. Clearly, in this case, they must be open circuit ports. Divide the network ports into three classes: P_{vo} (open circuit ports for which the voltages are not equal to zero), P_v' (open circuit ports for which the voltages are equal to zero) and P_v'' (the remaining ports which are Ohm's resistors and short circuits). Note that $P_{vo} \cup P_v'$ contains only open circuit ports and that $P_v' \cup P_v''$ contains all ports for which the voltage is equal to zero. Suppose that every port k in P_{vo} is oriented such that the $v_k > 0$. Let G be a partially oriented graph associated with the network in which the edges that correspond to ports in P_{vo} are oriented in accordance with the actual voltage orientations and the remaining edges are not oriented. According to the Painting Theorem, each edge in P_{vo} either belongs to a uniform cutset made of edges in $P_{vo} \cup P_v'$ only or to a uniform circuit made of edges in $P_{vo} \cup P_v''$ only. But every cutset made of edges in $P_{vo} \cup P_v'$ only is a cutset made of open circuit ports. This contradicts condition (i). Hence, every edge k in P_{vo} belongs to a uniform circuit made of edges in $P_{vo} \cup P_v''$ only. Since the circuit is uniform, in the associated Kirchhoff's voltage law all members appear with a positive sign. On the other hand every member of the sum is either positive (when it is related to the ports in P_{vo}) or is zero. According to our assumption P_{vo} is nonempty and hence at least one member of the sum is nonzero. This is a contradiction. Thus we conclude that P_{vo} is empty, that is all open circuits belong to P_v'. This completes the proof. □

Because every negative resistive one-port can be modelled by a positive resistive one-port and one nullator-norator pair (see Figure 1.27 of Section

1.8, Chapter 1), a general affine, time-invariant, resistive network can be described as a network consisting of positive Ohm's resistors, independent voltage and current sources and an equal number of nullators and norators.

Consider now a network that consists of Ohm's resistors, independent voltage and current sources and ideal 2-port transformers only. For such a network the following condition is necessary for unique solvability:

(ii) *there exists a tree containing all the voltage sources, none of the current sources and exactly one port for every ideal 2-port transformer.*

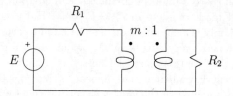

Figure 4.17. A network illustrating that the converse of (ii) generally is not true.

The converse is generally not true. Consider the network of Figure 4.17 in which all parameters are nonzero. If the relation $R_1 = -m^2 R_2$ holds, then we do not have a unique solution although condition (ii) holds. Otherwise, if $R_1 \neq -m^2 R_2$, the network has a unique solution.

Next, consider a network that consists of Ohm's resistors, independent voltage and current sources and gyrators only. For such a network the following condition is necessary for unique solvability:

(iii) *there exists a tree containing all the voltage sources, none of the current sources and either both or none of the ports for every gyrator.*

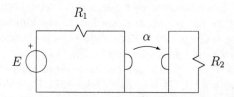

Figure 4.18. A network illustrating that the converse of (iii) generally is not true.

The converse is generally not true. Consider the network of Figure 4.18 in which all parameters are nonzero. If the relation $R_1 R_2 = -\alpha^2$ holds, then it has no unique solution although condition (iii) holds. Otherwise, if $R_1 R_2 \neq -\alpha^2$, the network has a unique solution. Note that because every ideal transformer can be represented as a cascade connection of two gyrators (see Figure 1.26 of Section 1.8, Chapter 1), networks with ideal transformers can be treated as special cases of networks with gyrators.

A tree that satisfies conditions (i), (ii) and (iii), is called a *normal tree*.

Suppose a network contains Ohm's resistors, independent voltage and current sources and an equal number of nullators and norators (recall that this is the general situation of time-invariant, affine resistive networks). For such networks the following condition is necessary for unique solvability:

(iv) *there is a pair of trees such that one includes all nullators and no norators, the other includes all norators and no nullators, both include all independent voltage sources and neither includes independent current sources.*

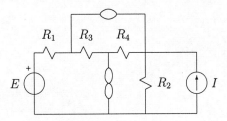

Figure 4.19. A network illustrating that the converse of (iv) is generally not true.

The converse is generally not true. Consider the network of Figure 4.19 in which all parameters are nonzero. If the relation $R_1R_4 = R_2R_3$ holds for this network, then it does not have a unique solution although condition (iv) holds. Otherwise, if $R_1R_4 \neq R_2R_3$, the network has a unique solution. If Ohm's resistors are absent, that is, if the network consists of independent voltage sources, independent current sources and an equal number of nullators and norators only, then condition (iv) is both necessary and sufficient for the network to have a unique solution. A pair of trees that satisfies condition (iv) is called a *conjugate* pair of trees.

4.6 Bibliographic notes

The notion of distance between trees was introduced by Watanabe [71], and later modified by Hakimi [25]. The concept of a central tree of a graph and the statement of Proposition 4.8 was originally proposed by Deo [15], and subsequently appeared in several papers. The idea of an extremal tree was proposed by Ohtsuki, Ishizaki and Watanabe [57] as an auxiliary notion in certain theorem proofs. The concepts of perfect and superperfect pairs of trees as well as the concept of basoid were introduced by Novak and Gibbons [52, 54, 55, 56]. A theorem with a similar formulation to that of Proposition 4.1, but involving the notion of diameter which is an entirely different notion of that introduced in this chapter, was presented by Harary, Mokken and Plantholt in [27]. On noncontradictory abstract affine network see also [48]. The proof of Proposition 4.32 mainly follows [30]. In many applications of lumped circuit theory, pairs of trees of a graph play an important role (see for example, Hasler [29], Milic and Novak [45] and Ohtsuki and Watanabe [58]). For further reading about graph and matroid oriented criteria for unique solvability of networks see [1], [5], [43], [17], [30] and [61].

5

Maximally Distant Pairs of Trees

In this chapter we consider the concept of maximally distant pairs of trees and the related concepts of principal minor and principal partition of a graph. We also characterize their relationship and relate these concepts to basoids and hybrid rank of a graph. Maximally distant pairs of trees of a graph together with principal minor, principal partition and hybrid rank inherently belong to hybrid oriented graph theory. These concepts have proved to be of relevance within a number of areas of application, including the solution of Shannon's famous switching game.

We refer to the following concepts introduced in Chapter 3 and Chapter 4: double independent subset, basoid, span, perfect pair of trees, superperfect pair of trees and diameter of a tree. We shall also occasionally refer to the definitions and theorems gathered together at the beginning of Chapter 3. The material of this chapter is general in the sense that it can be easily extended from graphs to matroids.

5.1 Preliminaries

A pair of trees (t_1, t_2) is said to be a *maximally distant pair of trees* if $|t_1 \backslash t_2| \geq |t' \backslash t''|$ for every pair of trees (t', t'') in G. As an example consider the graph of Figure 5.1 with two trees indicated by bold edges. Each of these trees is the complement of the other and therefore they obviously form a maximally distant pair of trees.

Because of the obvious fact that $|t_1 \backslash t_2|$ *is a maximum iff* $|E \backslash (t_1 \cup t_2)|$ *is a minimum,* the following statement is also true: *a pair of trees* (t_1, t_2) *of a graph G with edge set E is a maximally distant pair of trees iff the relation* $|E \backslash (t'_1 \cup t'_2)| \geq |E \backslash (t_1 \cup t_2)|$ *holds for all pairs of trees* (t'_1, t'_2) *of G.*

Proposition 5.1 *A maximally distant pair is a perfect pair.*

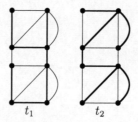

Figure 5.1. Two copies of the same graph with a pair of complementary trees indicated.

Proof If (t_1, t_2) is a maximally distant pair then $|t_1 \backslash t_2| \geq |t' \backslash t''|$ for every pair (t', t'') in the graph. For the particular case: $(t' = t_1, t'' = t_2)$ we have $|t_1 \backslash t_2| \geq |t_1 \backslash t''|$ for every t'', that is, t_2 is maximally distant from t_1. Also, for the case $(t' = t_1, t'' = t'')$, $|t_2 \backslash t_1| \geq |t_2 \backslash t'|$ for every t', that is, t_1 is maximally distant from t_2. Consequently, the pair (t_1, t_2) is a perfect pair.   □

Remark 5.2 The converse of Proposition 5.1 is generally not true, that is, *a perfect pair of trees need not be a maximally distant pair of trees.*

To see, consider Figure 5.2 which shows a perfect pairs of trees (t_1', t_2'), with distance of 5. This pair is not a maximally distant pair because for the same graph there is a pair with distance 7, as Figure 5.1 shows.

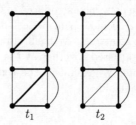

Figure 5.2. A perfect pair of trees with distance 5.

Proposition 5.3 *If (t_1, t_2) is a maximally distant pair of trees then both $t_1 \backslash t_2$ and $t_2 \backslash t_1$ are basoids.*

Proof Suppose that one of the subsets $t_1 \backslash t_2$ or $t_2 \backslash t_1$ is not a basoid, for example the subset $t_1 \backslash t_2$. Then there exists a basoid b that contains $t_1 \backslash t_2$ as a proper subset. Because every basoid is double independent, according to Lemma 4.19 of Chapter 4, there is a pair of trees (t_1', t_2') such that $t_1' \backslash t_2' = b$. We conclude that $|t_1 \backslash t_2| < |b| = |t_1' \backslash t_2'|$ which contradicts the assumption that (t_1, t_2) is a maximally distant pair of trees.   □

Remark 5.4 The converse of Proposition 5.3 is generally not true. That is, *if for a given pair of trees (t_1, t_2), both $t_1 \backslash t_2$ and $t_2 \backslash t_1$ are basoids, then (t_1, t_2) is not necessarily a maximally distant pair of trees.*

In order to see this consider Figure 5.3. This figure shows four copies of the same graph with different subsets of edges indicated with bold lines. Now $t_1 \backslash t_2$ and $t_2 \backslash t_1$ are both basoids but (t_1, t_2) is not a maximally distant pair of trees.

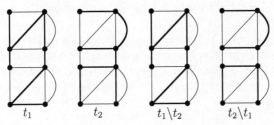

$$t_1 \qquad\qquad t_2 \qquad\qquad t_1 \backslash t_2 \qquad\qquad t_2 \backslash t_1$$

Figure 5.3. Four copies of the same graph with a pair of trees and their set differences indicated.

As an immediate consequence of Proposition 5.3 we have the following two corollaries:

Corollary 5.5 [to Proposition 5.3] *The cardinality of a largest basoid of a graph is equal to the maximal tree diameter in the graph.*

Denote by D the set of all tree diameters of a graph G and let B be the set of all basoid cardinalities of G. Then, the statement of Corollary 5.5 can be reformulated thus: $\max B = \max D$.

Corollary 5.6 [to Proposition 5.3] *A maximally distant pair of trees of a graph is a superperfect pair of trees of the graph.*

The following proposition provides a useful observation.

Proposition 5.7 *A pair (t_1, t_2) is a maximally distant pair of trees of a graph G iff the pair (t_1^*, t_2^*) is a maximally distant pair of cotrees of G.*

Proof The relation $|t_1| + |t_2| + |t_1^* \cap t_2^*| - |t_1 \cap t_2| = |E|$ obviously holds for any pair of trees (t_1, t_2) of G. Because $|t_1| + |t_2|$ is a constant for G, the minimum value of $|t_1^* \cap t_2^*|$ is attained for precisely those pairs (t_1, t_2) for which the minimum of $|t_1 \cap t_2|$ is attained and so the result follows. $\square$

The next proposition gives necessary and sufficient conditions for a subset of the edges of a graph to be a basoid of maximal cardinality.

Proposition 5.8 *A subset of edges b of a graph is a basoid of maximal cardinality (a dyad) iff b is a set difference of a maximally distant pair of trees.*

Proof Let b be a basoid of maximal cardinality. Because every basoid is double independent, there exists a pair of trees (t_1, t_2) such that $b = t_1 \backslash t_2$ (Lemma 4.19, Chapter 4). Suppose now that (t_1, t_2) is not a maximally distant pair of trees. Then for a maximally distant pair (t_1', t_2'), according to Proposition 5.3, the set difference $t_1' \backslash t_2' = b_1$ is also a basoid. But $|b_1| = |t_1' \backslash t_2'| > |t_1 \backslash t_2| = |b|$ which contradicts the assumption that b is a basoid of maximal cardinality.

Conversely, suppose that b is a set difference of a maximally distant pair of trees (t_1, t_2). Then according to Proposition 5.3, b is a basoid. Any other basoid also is a set difference of a perfect pair of trees (Lemma 4.19, Chapter 4). But $|t_1 \backslash t_2| \geq |t_1' \backslash t_2'|$ for all t_1', t_2' in the graph and consequently b is of maximal cardinality.   □

Remark 5.9 We have established that the following sequence of implications holds: *Maximally distant pair $\Rightarrow$ Superperfect pair $\Rightarrow$ Perfect pair.*

It is easy to see that a *disjoint pair of trees* is also a maximally distant pair and therefore a perfect pair. If $t_1^* = t_2$ and $t_2^* = t_1$ then (t_1, t_2) is called a *complementary pair of trees*. Figure 5.4 shows two copies of the same graph and the trees indicated are a complementary pair. The following implications obviously hold: *Complementary pair $\Rightarrow$ Disjoint pair $\Rightarrow$ Maximally distant pair.*

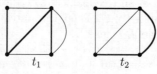

$$t_1 \qquad\qquad t_2$$

Figure 5.4. Two copies of the same graph with a pair of complementary trees indicated.

Remark 5.10 In general, concerning a maximally distant pair of trees (t_1, t_2) we can consider four cases:
  (a) $t_1 \cap t_2 = \emptyset$ and $t_1^* \cap t_2^* \neq \emptyset$.
  (b) $t_1 \cap t_2 = \emptyset$ and $t_1^* \cap t_2^* = \emptyset$.
  (c) $t_1 \cap t_2 \neq \emptyset$ and $t_1^* \cap t_2^* = \emptyset$.
  (d) $t_1 \cap t_2 \neq \emptyset$ and $t_1^* \cap t_2^* \neq \emptyset$.

Notice that if (a) holds then (t_1, t_2) is a disjoint pair of trees, if (b) holds (t_1, t_2) a complementary pair of trees and if (c) holds (t_1, t_2) a disjoint pair of cotrees. If a given pair of trees (t_1, t_2) satisfies any of (a), (b) or (c) then it follows immediately that the pair (t_1, t_2) is maximally distant. In cases (a) and (b) this is obvious. In case (c), this is a consequence of Proposition 5.7. However, if (d) holds, then t_1 and t_2 may or may not be maximally distant. Only rarely do a pair of trees of a graph satisfy (a), (b) or (c). The next two sections are devoted to maximally distant pairs of trees that satisfy (d).

5.2   Minor with respect to a pair of trees

For a given pair of trees (t_1, t_2) of a graph G let $t_1^* \cap t_2^* = Q$, and $t_1 \cap t_2 = P$.

An edge subset M of a graph G is said to be a *minor* (t_1, t_2) *of* G if the following conditions hold:

(M1) $Q \subseteq M$, and

(M2) $sp_c((t_1 \setminus t_2) \cap M) = M = sp_c((t_2 \setminus t_1) \cap M)$.

Dually, an edge subset N of a graph G is said to be a (*cominor*) *with respect to a pair of trees* (t_1, t_2) *of* G if the following conditions hold:

(N1) $P \subseteq N$, and

(N2) $sp_s((t_1 \setminus t_2) \cap N) = N = sp_s((t_2 \setminus t_1) \cap N)$.

Since obviously $t_2^* \setminus t_1^* = t_1 \setminus t_2$ and $t_1^* \setminus t_2^* = t_2 \setminus t_1$, it follows that condition (M2) for minors can be replaced with:

(M2)$'$ $sp_c((t_1^* \setminus t_2^*) \cap M) = M = sp_c((t_2^* \setminus t_1^*) \cap M)$.

Dually, condition (N2) for cominors can be replaced with:

(N2)$'$ $sp_s((t_1^* \setminus t_2^*) \cap N) = N = sp_s((t_2^* \setminus t_1^*) \cap N)$.

Consequently, we can also say that M (N) is a minor (cominor) with respect to a pair of cotrees (t_1^*, t_2^*).

Lemma 5.11 *Let M (N) be an arbitrary edge subset of G and let (t_1, t_2) be a pair of trees of G. If the relation $sp_c((t_1 \setminus t_2) \cap M) \supseteq M$ ($sp_s((t_1 \setminus t_2) \cap N) \supseteq N$) holds then $P \cap M$ ($Q \cap N$) is empty.*

Proof An edge belongs to $sp_c((t_1 \setminus t_2) \cap M) \setminus (t_1 \setminus t_2) \cap M$ ($sp_s((t_1 \setminus t_2) \cap N) \setminus (t_1 \setminus t_2) \cap N$) iff it forms a circuit (cutset) with edges in $(t_1 \setminus t_2) \cap M$ ($(t_1 \setminus t_2) \cap N$). Since no edge in P (Q) forms a circuit (cutset) with edges in $(t_1 \setminus t_2) \cap M$ ($(t_1 \setminus t_2) \cap N$) only, we conclude that $P \cap sp_c((t_1 \setminus t_2) \cap M)$ ($Q \cap sp_s((t_1 \setminus t_2) \cap N)$) is empty. But $sp_c((t_1 \setminus t_2) \cap M) \supseteq M$ ($sp_s((t_1 \setminus t_2) \cap N) \supseteq N$) and hence $P \cap M$ ($Q \cap N$) is also empty. □

Notice that if M (N) is a minor (cominor) of a graph G with respect to a pair of trees (t_1, t_2) of G then $sp_c((t_1 \setminus t_2) \cap M) = M$ ($sp_s((t_1 \setminus t_2) \cap N) = N$) and hence $sp_c((t_1 \setminus t_2) \cap M) \supseteq M$ ($sp_s((t_1 \setminus t_2) \cap N) \supseteq N$). Therefore, according to Lemma 5.11, $P \cap M = \emptyset$ ($Q \cap N = \emptyset$).

Consider the graph of Figure 5.5 and the pair of trees $t_1 = \{a, c, d, f, h\}$ and $t_2 = \{a, b, e, g, i\}$. Here $P = t_1 \cap t_2 = \{a\}$ and $Q = t_1^* \cap t_2^* = \{j\}$. The edge sets $M_1 = \{d, e, h, i, j\}$ and $M_2 = \{d, e, f, g, h, i, j\}$ are two minors

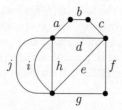

Figure 5.5. A graph having minors $M_1 = \{d, e, h, i, j\}$ and $M_2 = \{d, e, f, g, h, i, j\}$ of G with different cardinalities.

of G with different cardinalities, both with respect to the tree pair (t_1, t_2). This example illustrates the fact that there may be several minors associated with a pair of trees. Also, not every tree pair has an associated minor. For example there is no minor associated with the tree pair $t_1 = \{a, b, c, f, h\}$ and $t_2 = \{a, b, e, g, i\}$. As we show later in this section, for a given pair of trees there is a minor of G with respect to it iff it is a maximally distant pair of trees.

Lemma 5.12 *Let (t_1, t_2) be a pair of trees of a graph G and let K be an edge subset of G such that $Q \subseteq K$ and $P \subseteq K^*$, then:*
(E$_1$): $sp_c((t_1 \setminus t_2) \cap K) \supseteq K$ *iff* $sp_s((t_2 \setminus t_1) \cap K^*) \supseteq K^*$.
(E$_2$): $sp_c((t_2 \setminus t_1) \cap K) \supseteq K$ *iff* $sp_s((t_1 \setminus t_2) \cap K^*) \supseteq K^*$.

Proof (E$_1$): Notice that we assume $Q \subseteq K$ and $P \subseteq K^*$. Suppose that $sp_s((t_2 \setminus t_1) \cap K^*) \supseteq K^*$ is true and $sp_c((t_1 \setminus t_2) \cap K) \supseteq K$ is not. Then there is an edge $y_1 \in t_1^* \cap K$ such that the circuit C_1 that y_1 forms with edges in t_1 only, contains $x_1 \in t_1 \cap K^*$. Since every edge in t_1 forms a cutset with edges in t_1^* only, then x_1 does. Denote by S_1 a cutset that x_1 forms with edges in t_1^* only. According to Proposition 3.8, Chapter 3, applied to the partition (t_1, t_1^*), if $x_1 \in t_1 \cap K^*$ belongs to the circuit C_1, then y_1 belongs to a cutset S_1. Consequently, $sp_s((t_2 \setminus t_1) \cap K^*) \not\supseteq K^*$, which is a contradiction.

Conversely, suppose that $sp_c((t_1 \setminus t_2) \cap K) \supseteq K$ is true and $sp_s((t_2 \setminus t_1) \cap K^*) \supseteq K^*$ is not. Then there is an edge $x_2 \in t_1 \cap K^*$ such that a cutset S_2 that x_2 forms with edges in t_1^* only, contains $y_2 \in t_1^* \cap K$. Since every edge in t_1^* forms a circuit with edges in t_1 only, then y_2 does. Denote by C_2 a circuit that y_2 forms with edges in t_1 only. According to Proposition 3.8, Chapter 3, applied to the partition (t_1, t_1^*), if $y_2 \in t_1^* \cap K$ belongs to S_2, then $x_2 \in t_1 \cap K^*$ belongs to a circuit C_2. Consequently, $sp_c((t_1 \setminus t_2) \cap K) \not\supseteq K$, a contradiction.

(E$_2$): The proof is obtained from that for (E$_1$) by interchanging the roles of t_1 and t_2 □

Proposition 5.13 *Let (t_1, t_2) be a pair of trees of a graph G and let M be an edge subset of G. Then the following statements are equivalent:*

(1) M is a minor with respect to a pair of trees (t_1, t_2).

(1') $sp_c((t_1 \setminus t_2) \cap M) \supseteq M \subseteq sp_c((t_2 \setminus t_1) \cap M)$ and $Q \subseteq M$.

(1'') $t_1 \cap M$ and $t_2 \cap M$ are disjoint maximal circuit-less subsets of M and $Q \subseteq M$.

(2) M^* is a cominor with respect to a pair of trees (t_1, t_2).

(2') $sp_s((t_1 \setminus t_2) \cap M^*) \supseteq M^* \subseteq sp_s((t_2 \setminus t_1) \cap M^*)$ and $P \subseteq M^*$.

(2'') $t_1^* \cap M^*$ and $t_2^* \cap M^*$ are disjoint maximal cutset-less subsets of M^* and $P \subseteq M^*$.

(3) $sp_c((t_1 \setminus t_2) \cap M) = M$ and $sp_s((t_1 \setminus t_2) \cap M^*) = M^*$.

(3') $sp_c((t_1 \setminus t_2) \cap M) \supseteq M$ and $sp_s((t_1 \setminus t_2) \cap M^*) \supseteq M^*$.

(3'') $M \cap (t_1 \setminus t_2)$ is a maximal circuit-less subset of M and $M^* \cap (t_1 \setminus t_2)$ is a maximal cutset-less subset of M^*.

Proof (1) $\Rightarrow$ (1') Obvious.

(1') $\Rightarrow$ (1) Suppose (1') holds. We prove, by contradiction, that $sp_c((t_1 \setminus t_2) \cap M) \setminus M)$ and $sp_c((t_2 \setminus t_1) \cap M) \setminus M)$ are empty. Suppose there is an edge $x \in (sp_c((t_1 \setminus t_2) \cap M) \setminus M)$. According to Lemma 5.11, $sp_c((t_1 \setminus t_2) \cap M) \supseteq M$ implies that $P \cap M$ is empty. Hence, $sp_c((t_1 \setminus t_2) \cap M) \setminus M$ is a proper subset of $(t_2 \setminus t_1) \cap M^*$. Therefore, x belongs to $(t_2 \setminus t_1) \cap M^*$ and forms a circuit C_x with edges in $(t_1 \setminus t_2) \cap M$ only. From Lemma 5.12, $sp_c((t_2 \setminus t_1) \cap M) \supseteq M$ implies $sp_s((t_1 \setminus t_2) \cap M^*) \supseteq M^*$ and therefore x also forms a cutset S_x with edges in $(t_1 \setminus t_2) \cap M^*$ only. Since $(t_1 \setminus t_2) \cap M$ and $(t_1 \setminus t_2) \cap M^*$ are disjoint, it follows that $C_x \cap S_x = \{x\}$, which contradicts the Orthogonality Theorem. By interchanging the roles of t_1 and t_2 we obtain that $sp_c((t_2 \setminus t_1) \cap M) \setminus M)$ is also empty.

(1') $\Rightarrow$ (1'') Obviously (1') implies that $(t_1 \setminus t_2) \cap M$ and $(t_2 \setminus t_1) \cap M$ are disjoint maximal circuit-less subsets of M. According to Lemma 5.11, $sp_c((t_1 \setminus t_2) \cap M) \supseteq M$ implies that $P \cap M$ is empty and consequently, $(t_1 \setminus t_2) \cap M = t_1 \cap M$ and $(t_2 \setminus t_1) \cap M = t_2 \cap M$, which completes the proof.

(1'') $\Rightarrow$ (1') If (1'') holds then $P \cap M$ is empty and hence $(t_1 \setminus t_2) \cap M = t_1 \cap M$ and $(t_2 \setminus t_1) \cap M = t_2 \cap M$. Consequently, $(t_1 \setminus t_2) \cap M$ and $(t_2 \setminus t_1) \cap M$ are maximal circuit-less subsets of M which implies $sp_c((t_1 \setminus t_2) \cap M) \supseteq M \subseteq sp_c((t_2 \setminus t_1) \cap M)$.

(2) $\Rightarrow$ (2') Obvious.
(2') $\Rightarrow$ (2'') By a proof dual to the proof of (1') $\Rightarrow$ (1'').
(2'') $\Rightarrow$ (2') By a proof dual to the proof of (1'') $\Rightarrow$ (1').
(3) $\Rightarrow$ (3') Obvious.

$(3') \Rightarrow (3)$ Suppose $(3')$ holds. According to Lemma 5.11, $(3')$ implies that $P \cap M$ and $Q \cap M^*$ are empty and hence $Q \subseteq M$ and $P \subseteq M^*$. Suppose there is an edge $x \in (sp_c((t_1 \backslash t_2) \cap M) \backslash M)$. Since $P \cap M$ is empty, $sp_c((t_1 \backslash t_2) \cap M) \backslash M$ is a proper subset of $(t_2 \backslash t_1) \cap M^*$. Then, x belongs to $(t_2 \backslash t_1) \cap M^*$ and forms a circuit C_x with edges in $(t_1 \backslash t_2) \cap M$ only. Because $sp_s((t_1 \backslash t_2) \cap M^*) \supseteq M^*$, x also forms a cutset S_x with edges in $(t_1 \backslash t_2) \cap M^*$ only. Since $(t_1 \backslash t_2) \cap M$ and $(t_1 \backslash t_2) \cap M^*$ are disjoint, it follows that $C_x \cap S_x = \{x\}$, which contradicts the Orthogonality Theorem.

Dually, suppose there is an edge $y \in (sp_s((t_1 \backslash t_2) \cap M^*) \backslash M^*)$. Since $Q \cap M^*$ is empty, $sp_s((t_1 \backslash t_2) \cap M^*) \backslash M^*$ is a proper subset of $(t_2 \backslash t_1) \cap M$. Clearly, y forms a cutset S_y with edges in $(t_1 \backslash t_2) \cap M^*$ only. Because $sp_c((t_1 \backslash t_2) \cap M) \supseteq M$, y also forms a circuit C_y with edges in $(t_1 \backslash t_2) \cap M$ only. Since $(t_1 \backslash t_2) \cap M$ and $(t_1 \backslash t_2) \cap M^*$ are disjoint, it follows that $C_y \cap S_y = \{y\}$, which contradicts the Orthogonality Theorem.

$(3') \Leftrightarrow (3'')$ Obvious.

$(1') \Leftrightarrow (3')$ Suppose $(1')$ holds. According to Lemma 5.11, $sp_c((t_1 \backslash t_2) \cap M) \supseteq M$ implies that $P \cap M$ is empty. Then, according to Lemma 5.12, $M \subseteq sp_c((t_2 \backslash t_1) \cap M)$ implies $sp_s((t_1 \backslash t_2) \cap M^*) \supseteq M^*$.

Conversely, if $(3')$ holds, according to Lemma 5.11, both $Q \cap M^*$ and $P \cap M$ are empty. According to Lemma 5.12, $sp_s((t_1 \backslash t_2) \cap M^*) \supseteq M^*$ implies $M \subseteq sp_c((t_2 \backslash t_1) \cap M)$, which completes the proof.

$(2') \Leftrightarrow (3')$ By a proof dual to the proof of $(1') \Leftrightarrow (3')$. $\square$

Using dual arguments we can prove also the following proposition.

Dual of Proposition 5.13 *Let (t_1, t_2) be a pair of trees of a graph G and let N be an edge subset of G. Then the following statements are equivalent:*

(1) *N is a cominor of G with respect to (t_1, t_2).*

$(1')$ $sp_s((t_1 \backslash t_2) \cap N) \supseteq N \subseteq sp_s((t_2 \backslash t_1) \cap N)$ *and* $P \subseteq N$.

$(1'')$ $t_1^* \cap N$ *and* $t_2^* \cap N$ *are disjoint maximal cutset-less subsets of N and* $P \subseteq N$.

(2) *N^* is a minor of G with respect to (t_1, t_2).*

$(2')$ $sp_c((t_1 \backslash t_2) \cap N^*) \supseteq N^* \subseteq sp_c(((t_2 \backslash t_1) \cap N) \cap N^*)$ *and* $Q \subseteq N^*$.

$(2'')$ $t_1 \cap N^*$ *and* $t_2 \cap N^*$ *are disjoint maximal circuit-less subsets of N^* and* $Q \subseteq N^*$.

(3) $sp_c((t_1 \backslash t_2) \cap N^*) = N^*$ *and* $sp_s((t_1 \backslash t_2) \cap N) = N$.

$(3')$ $sp_c((t_1 \backslash t_2) \cap N^*) \supseteq N^*$ *and* $sp_s((t_1 \backslash t_2) \cap N) \supseteq N$.

$(3'')$ $N^* \cap (t_1 \backslash t_2)$ *is a maximal circuit-less subset of N^* and* $N \cap (t_1 \backslash t_2)$ *is a maximal cutset-less subset of N.*

5.3 Principal sequence

Given a pair of trees (t_1, t_2) of a graph G construct the sequence A_Q^k, $k \geq 0$, where $A_Q^0 (= Q = t_1^* \cap t_2^*)$, A_Q^1 is the union of all fundamental circuits that edges in Q form with respect to t_1 and all fundamental circuits that edges in Q form with respect to t_2, A_Q^2 is the union of all fundamental circuits that edges in $A_Q^1 \setminus t_1$ form with edges in t_1 and all fundamental circuits that edges in $A_Q^1 \setminus t_2$ form with edges in t_2, etc., including A_Q^1. Inductively, for all $k \geq 1$, let A_Q^{k+1} be the union of all fundamental circuits that edges in $A_Q^k \setminus t_1$ form with edges in t_1 and all fundamental circuits that edges in $A_Q^k \setminus t_2$ form with edges in t_2, including A_Q^k. We call the sequence of edge subsets $A_Q^0, A_Q^1, \ldots, A_Q^k, \ldots$ a *principal sequence* of edge subsets associated with the pair (t_1, t_2) of the graph G. By construction, $A_Q^k \subseteq A_Q^{k+1}$, for all $k \geq 0$. Moreover, $Q \subseteq A_Q^k \setminus t_1$ and $Q \subseteq A_Q^k \setminus t_2$, for all $k \geq 0$. That is, the sequence $A_Q^0, A_Q^1, \ldots, A_Q^k, \ldots$ is a way of enlarging $Q = t_1^* \cap t_2^*$. Clearly, because the edge set is finite, there is an r such that, $A_Q^r = A_Q^k$ for all $k > r$. Thus, all members of the sequence for $k \geq r$ coincides with the same edge subset which we denote by A_Q. Clearly, A_Q is the largest member of the principal sequence of edge subsets associated with the pair (t_1, t_2). This edge subset is of crucial importance for this and the following section.

Remark 5.14 It is clear by construction that for $k \geq 1$ any edge in $(A_Q^{k+1} \setminus A_Q^k) \cap t_1$ belongs to a circuit that an edge in $(A_Q^k \setminus A_Q^{k-1}) \cap t_2$ forms with edges in t_1 only. Similarly, any each edge in $A_Q^{k+1} \setminus A_Q^k \cap t_2$ belongs to a circuit that an edge in $A_Q^k \setminus A_Q^{k-1} \cap t_1$ forms with edges in t_2 only. According to Proposition 3.8, Chapter 3, applied to the pair (t_1, t_1^*), it follows that the cutset that any edge in $(A_Q^{k+1} \setminus A_Q^k) \cap t_1$ forms with edges in t_1^* only has a nonempty intersection with $(A_Q^k \setminus A_Q^{k-1}) \cap t_2$. Similarly, the cutset that any edge in $(A_Q^{k+1} \setminus A_Q^k) \cap t$ forms with edges in t_2^* only has a nonempty intersection with $(A_Q^k \setminus A_Q^{k-1}) \cap t_1$.

Proposition 5.15 *Let A_Q be the largest subset of the principal sequence of edge subsets associated with a pair of trees (t_1, t_2) of a graph G. Then the following hold: (a) $A_Q \supseteq Q$. (b) If $P \cap A_Q$ is empty, then $sp_c((t_1 \setminus t_2) \cap A_Q) \supseteq A_Q \subseteq sp_c((t_2 \setminus t_1) \cap A_Q)$.*

Proof (a) follows immediately from construction of A_Q. If $P \cap A_Q$ is empty then every edge in $A_Q \setminus t_1$ forms a circuit with edges in $A_Q \cap t_1 = A_Q \cap (t_1 \setminus t_2)$ and every edge in $A_Q \setminus t_2$ forms a circuit with edges in $A_Q \cap t_2 = A_Q \cap (t_2 \setminus t_1)$. Consequently, $sp_c((t_1 \setminus t_2) \cap A_Q) \supseteq A_Q \subseteq sp_c((t_2 \setminus t_1) \cap A_Q)$, that is, (b) holds. $\square$

In dual manner, for a given pair of trees (t_1, t_2) of a graph G, we can construct the sequence $B_P^0, B_P^1, \ldots, B_P^k, \ldots$ which we call a *principal cosequence* associated with a pair of trees (t_1, t_2) of G. We can similarly define the set B_P. By dual arguments we also have:

Dual of Proposition 5.15 *Let B_P be the largest subset of the principal cosequence of edge subsets associated with a pair of trees (t_1, t_2) of a graph. Then the following hold:* (a) $B_P \supseteq P$ *and* (b) *if $Q \cap B_P$ is empty, then* $sp_s((t_1 \setminus t_2) \cap B_P) \supseteq B_P \subseteq sp_s((t_2 \setminus t_1) \cap B_P)$.

Proposition 5.16 *Let A_Q be the largest subset of the principal sequence of edge subsets associated with a pair of trees (t_1, t_2) of a graph G. Then, A_Q is a minor of G with respect to (t_1, t_2) iff $A_Q \cap P$ is empty.*

Proof Suppose that A_Q is a minor of G with respect to (t_1, t_2). Then, according to Proposition 5.13, A_Q^* is a cominor of G with respect to (t_1, t_2) and hence, $P \subseteq A_Q^*$. Therefore, $A_Q \cap P$ is empty.

Conversely, if the intersection $A_Q \cap P$ is empty, then, from Propositions 5.15 and 5.13, A_Q is a minor with respect to a pair of trees. $\square$

Dual of Proposition 5.16 *Let B_P be the largest subset of the principal cosequence of edge subsets associated with a pair of trees (t_1, t_2) of a graph. Then B_P is a cominor of G with respect to (t_1, t_2) iff $B_P \cap Q$ is empty.*

Lemma 5.17 *Given a pair of trees (t_1, t_2) of a graph G, suppose $e_0 \in Q$ and let $e_1 \in t_1$ be an edge that belongs to the unique circuit that e_0 forms with edges in t_1 only. Let $t_1' = t_1 \setminus \{e_1\}) \cup \{e_0\} = t_1 \oplus \{e_1, e_0\}$ and $t_2' = t_2$. If $e_1 \notin P$ then $P' = t_1' \cap t_2' = t_1 \cap t_2 = P$ and (t_1', t_2') has distance equal to that of the pair (t_1, t_2). If however $e_1 \in P$ then $P' \subset P$ and (t_1', t_2') has distance one greater than that of the pair (t_1, t_2).*

Proof Clearly, $e_0 \notin P$ because Q and P are disjoint. If $e_1 \notin P$ then obviously after replacing t_1 with $t_1' = t_1 \setminus \{e_1\}) \cup \{e_0\}$ and keeping t_2 unchanged ($t_2' = t_2$) $t_1' \cap t_2' = t_1 \cap t_2$, that is, $P' = P$ and therefore (t_1', t_2') has distance equal to that of the pair (t_1, t_2). If however $e_1 \in P$ then since, $e_0 \notin t_2$, it follows that $P \subset P'$ and (t_1', t_2') has distance one greater than that of the pair (t_1, t_2). $\square$

Proposition 5.18 *Let (t_1, t_2) be a maximally distant pair of trees of a graph and let A_Q be the largest subset of the principal sequence of edge subsets associated with (t_1, t_2). Then $A_Q \cap P = \emptyset$.*

Proof The proof is by contradiction. Let us assume that $A_Q \cap P$ is not empty. From the construction of the sequence $A_Q^0 = Q, A_Q^1, \dots, A_Q^k, \dots$ it follows that there is a positive integer s such that the intersection $P \cap A_Q^s$ is empty and $P \cap A_Q^{s+1}$ is not. Let $e_s \in A_Q^s$ and let $e_{s+1} \in A_Q^{s+1} \cap P$. According to Remark 5.14, we can construct a sequence of edges $e_0, e_1, \dots, e_s, e_{s+1}$, associated with a maximally distant pair of trees (t_1, t_2) , by backtracking from e_{s+1}, such that $e_0 \in A_Q^0 = Q$, $e_{s+1} \in A_Q^{s+1} \cap P$ and for even $k \leq s$, $e_{k+1} \in (A_Q^{k+1} \setminus A_Q^k) \cap t_1$ and for odd $k \leq s$, $e_{k+1} \in (A_Q^{k+1} \setminus A_Q^k) \cap t_2$.

It is clear by construction of $A_Q^0 = Q, A_Q^1, \ldots, A_Q^k, \ldots$ that for the sequence of edges $e_0, e_1, \ldots, e_s, e_{s+1}$, which is associated with a maximally distant pair of trees (t_1, t_2), the following conditions hold:

(i) $e_0 \in Q$

(ii) $e_{s+1} \in P$

(iii) for $0 \leq k \leq s$, e_{k+1} belongs to the fundamental circuit C_k that e_k forms either with respect to the tree t_1, if k is even or with respect to the tree t_2, if k is odd.

(iv) for $0 \leq k < s$, $C_k \cap P$ is the empty set but $C_s \cap P$ is not (because it contains e_{s+1}).

(v) for $0 \leq p < k \leq s$, $e_{k+2} \notin C_p$.

We consider three cases: $s = 0$, $s = 1$ and $s > 1$. Let $s = 0$ and suppose the sequence e_0, e_1 satisfies conditions (i) to (v) with respect to the pair of trees (t_1, t_2). For this case conditions (i) to (v) reduce to $e_0 \in Q$ and $e_1 \in C_0 \cap P$. Let $t_1' = t_2$ and $t_2' = t_1 \cup \{e_0\} \setminus \{e_1\}$. From Lemma 5.17, (t_1', t_2') is a new pair of trees with distance one greater than the original pair. This contradicts the assumption that (t_1, t_2) is a maximally distant pair of trees.

Let $s = 1$ and suppose the sequence e_0, e_1, e_2 satisfies conditions (i) to (v) with respect to the pair of trees (t_1, t_2) in terms of fundamental circuits C_k, $0 \leq k \leq s$ and sets $Q = t_1^* \cap t_2^*$ and $P = t_1 \cap t_2$. Conditions (i) to (v) then reduce to $e_0 \in Q$, $e_1 \in C_0 \cap (t_1 \setminus t_2)$, $e_2 \in C_1 \cap P$, $e_2 \notin C_0$ and $C_0 \cap P$ is empty. Let $t_1^1 = t_2$ and $t_2^1 = t_1 \cup \{e_0\} \setminus \{e_1\}$. Since $e_1 \notin P$, we see from Lemma 5.17 that (t_1^1, t_2^1) is a pair of trees with distance equal to that of the pair (t_1, t_2). The sequence e_1, e_2 also satisfies conditions (i) to (v) with respect to the pair (t_1^1, t_2^1). Let $t_1^2 = t_2^1$ and $t_2^2 = t_1^1 \cup \{e_1\} \setminus \{e_2\}$. According to Lemma 5.17, (t_1^2, t_2^2) is a pair of trees with distance one greater than the pair (t_1^1, t_2^1) and consequently, one greater than (t_1, t_2). This contradicts the assumption that (t_1, t_2) is a maximally distant pair of trees.

Let $s > 1$ and suppose the sequence $e_0, e_1, \ldots, e_s, e_{s+1}$ satisfies conditions (i) to (v) with respect to the pair of trees (t_1, t_2) in terms of fundamental circuits C_k, $0 \leq k \leq s$ and sets $Q = t_1^* \cap t_2^*$ and $P = t_1 \cap t_2$. Let $t_1^1 = t_2$ and $t_2^1 = t_1 \cup \{e_0\} \setminus \{e_1\}$. As $s \neq 0$, it follows from condition (iv) that the circuit $C_0 \cap P$ is empty. According to Lemma 5.17, (t_1^1, t_2^1) is also a pair of trees with the distance equal to that of the pair (t_1, t_2). For $1 \leq k \leq s$, denote by C_k^1 the fundamental circuit defined by the edge e_k with respect to the tree t_1^1 if $k + 1$ is even and with respect to the tree t_2^1 if $k + 1$ is odd. Clearly, for even k, $1 \leq k \leq s$, $C_k^1 = C_k$ if C_k does not contain e_1 and $C_k^1 = C_k \oplus C_0$ if C_k contains e_1. This follows immediately from the obvious fact that C_0, which is the fundamental circuit that e_0 forms with edges in t_1 only, is also the fundamental circuit that e_1 forms with edges in t_2^1 only. For odd k, $1 \leq k \leq s$,

$C_k^1 = C_k$ always. Let $Q^1 = (t_1^1)^* \cap (t_2^1)^*$ and $P^1 = t_1^1 \cap t_2^1$. Then the sequence $e_1, \ldots, e_s, e_{s+1}$, in terms of fundamental circuits C_k^1 and sets Q^1 and P^1 also satisfies conditions (i) to (v) with respect to the tree pair (t_1^1, t_2^1).

Inductively, suppose that the sequence $e_p, e_{p+1}, \ldots, e_s, e_{s+1}$ satisfies conditions (i) to (v) with respect to a pair of trees (t_1^p, t_2^p) in terms of fundamental circuits C_k^p, $p \le k \le s$, and sets $Q^p = (t_1^p)^* \cap (t_2^p)^*$ and $P^p = t_1^p \cap t_2^p$. Here C_k^p denotes the circuit defined by the edges e_k with respect to the tree t_1^p if $k + p$ is even and with respect to the tree t_2^p if $k+p$ is odd. Notice that e_{k+1} belongs to the fundamental circuit C_k^p. Let $t_1^{p+1} = t_2^p$ and $t_2^{p+1} = t_1^p \cup \{e_p\} \setminus \{e_{p+1}\}$. If $s \ne p$, it follows from condition (iv) that the circuit $C_p^p \cap P$ is empty. From Lemma 5.17, (t_1^{p+1}, t_2^{p+1}) is a pair of trees with distance equal to that of the pair (t_1^p, t_2^p). Let C_k^{p+1} be the fundamental circuit defined for $p+1 \le k \le s$, by an edge e_k with respect to t_1^{p+1} if $k+p+1$ is even that is, with respect to t_2^{p+1} if $k+p+1$ is odd. Clearly, for even k, $p+1 \le k \le s$, $C_k^{p+1} = C_k^p$, if C_k^p does not contain e_p, and $C_k^{p+1} = C_k^p \oplus C_p^p$ if C_k^p contains e_p. This follows immediately from the obvious fact that C_p^p, which is the fundamental circuit that e_p forms with edges in t_1^p only, is at the same time the fundamental circuit that e_{p+1} forms with edges in t_1^{p+1} only. For odd k, $p + 1 \le k \le s$, $C_k^{p+1} = C_k^p$, always. Let $Q^{p+1} = (t_1^{p+1})^* \cap (t_2^{p+1})^*$ and $P^{p+1} = t_1^{p+1} \cap t_2^{p+1}$. Then, the sequence $e_{P+1}, \ldots, e_s, e_{s+1}$, in terms of circuits C_k^{p+1} and the sets Q^{p+1} and P^{p+1}, also satisfies conditions (i) to (v) with respect to the pair (t_1^{p+1}, t_2^{p+1}).

After s repetitions of the same procedure we obtain a tree pair (t_1^s, t_2^s) with the same distance as the original pair (t_1, t_2) and the corresponding sets Q^s and P^s. Let $C_k^{s+1} = C_k^s$ if C_k^s does not contain e_s and $C_k^{s+1} = C_k^s \oplus C_k^s$ if C_k^s contains e_s. The sequence e_s, e_{s+1} satisfies conditions (i) to (v) with respect to the tree pair (t_1^s, t_2^s) providing that $e_s \in Q^s$ and $e_{s+1} \in C_s \cap P^s$. Let $t_1^{s+1} = t_2^s$ and $t_2^{s+1} = t_1^s \cup \{e_s\} \setminus \{e_{s+1}\}$. According to Lemma 5.17, (t_1^{s+1}, t_2^{s+1}) is a new pair of trees of distance one greater than the pair (t_1^s, t_2^s). But the pair (t_1^s, t_2^s) has the same distance as the pair (t_1, t_2). Consequently, (t_1^{s+1}, t_2^{s+1}) has distance one greater than (t_1, t_2). This contradicts the assumption that (t_1, t_2) is a maximally distant pair of trees. $\qquad \square$

An augmentation of a set $P = t_1 \cap t_2$ (as compared to the set $Q = t_1^* \cap t_2^*$) leads to the generation of a sequence $B_P^0, B_P^1, \ldots, B_P^j, \ldots$ (as compared to $A_Q^0, A_Q^1, \ldots, A_Q^k, \ldots$). Here $B_P^0 = t_1 \cap t_2$ and $B_P^{j+1}, j = 1, 2, \ldots$ is the union of all cutsets that edges in $B_P^j \setminus t_1$ form with edges in t_1^* and all cutsets that edges in $B_P^j \setminus t_2^*$ form with edges in t_2^* including B_P^j. Obviously, $B_P^j \subseteq B_P^{j+1}$. Clearly (because the edge set is finite), for some r, $B_P^r = B_P^j$ for all $j > r$. For such an r by analogy with the previous material we write $B_P = B_P^r$.

Dual of Proposition 5.18 *Let (t_1, t_2) be a maximally distant pair of trees of a graph and let B_P be the largest subset of the principal cosequence of edge subsets associated with (t_1, t_2). Then $B_P \cap Q$ is empty.*

If (t_1, t_2) is a disjoint pair of trees and/or (t_1^*, t_2^*) is a disjoint pair of cotrees then (t_1, t_2) is a maximally distant pair of trees. If however both $t_1 \cap t_2 \neq \emptyset$ and $t_1^* \cap t_2^* \neq \emptyset$, then (t_1, t_2) may not be a maximally distant pair of trees.

Lemma 5.19 *Let M be a minor of a graph G with respect to a pair of trees (t_1, t_2) of G. Then for any pair of trees (t_a, t_b):*

$$|t_a^* \cap t_b^*| \geq |M| - |(t_a \cap M) \cup (t_b \cap M)| \geq |M| - |(t_1 \cap M) \cup (t_2 \cap M)| = |t_1^* \cap t_2^*|.$$

Proof For any edge subset M of G and for any pair of trees (t_a, t_b) of G we have:

$$t_a^* \cap t_b^* \supset M \cap t_a^* \cap t_b^* = M \setminus ((t_a \cap M) \cup (t_b \cap M))$$

and

$$|M \setminus ((t_a \cap M) \cup (t_b \cap M))| = |M| - |(t_a \cap M) \cup (t_b \cap M)|.$$

Hence, for any subset M and for any pair of trees (t_a, t_b),

$$|t_a^* \cap t_b^*| \geq |M| - |(t_a \cap M) \cup (t_b \cap M)|. \tag{1}$$

Also, for any subset M and for any pair of trees (t_a, t_b),

$$|(t_a \cap M)| + |(t_b \cap M)| \geq |(t_a \cap M) \cup (t_b \cap M)|$$

where equality is attained iff $t_a \cap M$ and $t_b \cap M$ are disjoint.

Let M be a minor of G with respect to the pair (t_1, t_2). Then, from Lemma 5.11, $P \cap M$ is empty and hence $t_1 \cap M$ and $t_2 \cap M$ are disjoint. Therefore,

$$|(t_1 \cap M) \cup (t_2 \cap M)| = |(t_1 \cap M)| + |(t_2 \cap M)|.$$

Since M is a minor with respect to (t_1, t_2), according to implication $(1) \Rightarrow (1'')$ of Proposition 5.13, $t_1 \cap M$ and $t_2 \cap M$ are maximal circuit-less subsets of M. Since $t_a \cap M$ and $t_b \cap M$ are also circuit-less subsets of M:

$$|t_1 \cap M| \geq |t_a \cap M|$$

and

$$|t_2 \cap M| \geq |t_b \cap M|.$$

Accordingly,

$$
\begin{aligned}
|(t_1 \cap M) \cup (t_2 \cap M)| &= |(t_1 \cap M)| + |(t_2 \cap M)| \\
&\geq |(t_a \cap M)| + |(t_b \cap M)| \geq |(t_a \cap M) \cup (t_b \cap M)|.
\end{aligned}
$$

Consequently,

$$|M| - |(t_a \cap M) \cup (t_b \cap M)| \geq |M| - |(t_1 \cap M) \cup (t_2 \cap M)|. \tag{2}$$

On the other hand since M is a minor with respect to (t_1, t_2) it follows that $Q = t_1^* \cap t_2^* \subseteq M$ and hence:

$$|t_1^* \cap t_2^*| = M \cap t_1^* \cap t_2^* = |M| - |(t_1 \cap M) \cup (t_2 \cap M)|. \qquad (3)$$

From (1), (2), and (3) we finally obtain

$$\begin{aligned} |t_a^* \cap t_b^*| &\geq |M| - |(t_a \cap M) \cup (t_b \cap M)| \geq |M| - |(t_1 \cap M) \cup (t_2 \cap M)| \\ &= |t_1^* \cap t_2^*|. \end{aligned} \qquad \square$$

Proposition 5.20 *Let (t_1, t_2) be a maximally distant pair of trees of a graph and let A_Q be the largest subset of the principal sequence of edge subsets associated with (t_1, t_2) and let B_P be the largest subset of the principal cosequence of edge subsets associated with (t_1, t_2). Then the following statements are equivalent:*

(1) *(t_1, t_2) is a maximally distant pair of trees of a graph G;*

(2) *there is a minor of G with respect to the pair (t_1, t_2);*

(3) *$A_Q \cap P$ is empty;*

(4) *$A_Q \cap B_P$ is empty.*

Proof $(1) \Rightarrow (2)$ Let (t_1, t_2) be a maximally distant pair of trees and let $Q = t_1^* \cap t_2^*$. Then, from Proposition 5.16, $P = t_1 \cap t_2$ and A_Q are disjoint which implies (Proposition 5.15), that A_Q is a minor of G with respect to the pair (t_1, t_2). Setting $M = A_Q$ we have proved the existence of a subset M.

$(2) \Rightarrow (1)$ Let (t_1, t_2) be a pair of trees of G and suppose that there exists a minor M of G with respect to the pair (t_1, t_2). Choose another pair (t_a, t_b) of G. According to Lemma 5.19, $|t_a^* \cap t_b^*| \geq |t_1^* \cap t_2^*|$. But, according to Proposition 5.7, $|t_a^* \cap t_b^*| \geq |t_1^* \cap t_2^*|$ iff $|t_a \cap t_b| \leq |t_1 \cap t_2|$. Since (t_a, t_b) is an arbitrary pair of trees of G we conclude that (t_1, t_2) is a maximally distant pair of trees of G.

$(1) \Rightarrow (3)$ Coincides with Proposition 5.18.

$(3) \Rightarrow (2)$ Immediately follows from Proposition 5.15.

$(3) \Leftrightarrow (4)$ Suppose $A_Q \cap P$ is empty. Then, according to Proposition 5.15, A_Q is a minor of G with respect to the tree pair (t_1, t_2). From Proposition 5.13 it follows that A_Q^* is a cominor with respect to the tree pair (t_1, t_2) and hence $B_P \subseteq A_Q^*$. Therefore $B_P \cap A_Q$ is empty.

Conversely, suppose that $A_Q \cap B_P$ is empty. Since $P \subseteq B_P$ it immediately follows that $A_Q \cap P$ is empty too. $\qquad \square$

Dual of Proposition 5.20 *Let (t_1, t_2) be a maximally distant pair of trees of a graph and let A_Q be the largest subset of the principal sequence of edge* *subsets associated with (t_1, t_2) and let B_P be the largest subset of the principal cosequence of edge subsets associated with (t_1, t_2). Then the following statements are equivalent:*

(1) *(t_1, t_2) is a maximally distant pair of trees of a graph G;*

(2) *there is a cominor of G with respect to the pair (t_1, t_2);*

(3) *$B_P \cap Q$ is empty;*

(4) *$B_P \cap A_Q$ is empty;*

According to Proposition 5.7 a cotree pair (t_1^*, t_2^*) of a graph G is a maximally distant pair of cotrees of a graph G iff the pair (t_1, t_2) is a maximally distant pair of trees of G. Hence the dual of Proposition 5.20 can be used not only for testing whether a pair of cotrees is maximally distant, but also to test if a pair of trees is maximally distant.

5.4   The principal minor

As an immediate consequence of Propositions 5.18 and 5.15 we have the following corollary:

Corollary 5.21 [to Propositions 5.18 and 5.15] *Let (t_1, t_2) be a maximally distant pair of trees of a graph and let A_Q be the largest subset of the principal sequence of edge subsets associated with (t_1, t_2). Then A_Q is a minor of G with respect to (t_1, t_2).*

In this section we prove that no proper subset of A_Q is a minor of G with respect to (t_1, t_2). Moreover we prove that A_Q is the unique minor of G.

We first consider the following proposition. Recall conditions (i) and (ii) of the definition of a minor of G with respect to a pair of trees of G.

Proposition 5.22 *Let (t_1, t_2) be a maximally distant pair of trees of a graph G. If M_1 and M_2 are minors of G with respect to the pair (t_1, t_2), then $M_1 \cap M_2$ is also a minor with respect to the pair (t_1, t_2).*

Proof Suppose that M_1 and M_2, are minors of G with respect to the maximally distant pair of trees (t_1, t_2). Then $t_1^* \cap t_2^* \subseteq M_1$ and $t_1^* \cap t_2^* \subseteq M_2$ which implies that $t_1^* \cap t_2^* \subseteq M_1 \cap M_2$. From the equivalence of statements (1) and (1″) of Proposition 5.13, $t_1 \cap M_1$ and $t_2 \cap M_1$ are disjoint circuit-less subsets of M_1. Because $M_1 \cap M_2 \subseteq M_1$, it follows that $t_1 \cap (M_1 \cap M_2)$ and $t_2 \cap (M_1 \cap M_2)$ are disjoint circuit-less subsets of $M_1 \cap M_2$. To show that $t_1 \cap (M_1 \cap M_2)$ and $t_2 \cap (M_1 \cap M_2)$ are maximal such subsets in $M_1 \cap M_2$, consider an edge $e \in (M_1 \cap M_2) \setminus t_1$. Denote by $C(e, t_1)$ the fundamental circuit that e forms with edges in t_1 only. Recall that $t_1 \cap M_1$ is a maximally circuit-less subset of M_1 and that $t_1 \cap M_2$ is a maximally circuit-less subset of M_2. Also note that both $t_1 \cap M_1$ and $t_1 \cap M_2$ are subsets of t_1. Because of the uniqueness of $C(e, t_1)$, this circuit necessarily coincides with both the fundamental circuit that e may form with edges in $t_1 \cap M_1$ and the fundamental circuit that e forms with edges in $t_1 \cap M_2$. Hence, $C(e, t_1) \subseteq t_1 \cap (M_1 \cap M_2)$. That is, $t_1 \cap (M_1 \cap M_2)$ is a maximal circuit-less subset of $M_1 \cap M_2$. Similarly, $t_2 \cap (M_1 \cap M_2)$ is a maximal circuit-less subset of $M_1 \cap M_2$. From the equivalence of statements (1) and (1″) of Proposition 5.13, $M_1 \cap M_2$ is a minor with respect to the same pair (t_1, t_2).   □

We say that a minor M of G with respect to a pair (t_1, t_2) of trees of G is a *minimal minor* of G with respect to the tree pair (t_1, t_2) if no proper subset of M is also a minor of G with respect to the same tree pair (t_1, t_2).

As an immediate consequence of Proposition 5.22 we have the following proposition.

Proposition 5.23 *There is a unique minimal minor of a graph G with respect to a maximally distant pair of trees of G.*

Proof Consider the family of all minors of G with respect to a maximally distant pair of trees (t_1, t_2). Clearly, their intersection is nonempty, because every member of the family contains $t_1^* \cap t_2^*$ as a proper subset. According to Proposition 5.22, this intersection belongs to the family and also belongs as a proper subset to each member of the family. Hence, it is the unique minimal subset of this family.   □

Proposition 5.24 *Let (t_1, t_2) be a maximally distant pair of trees of a graph G and let A_Q be the largest subset of the principal sequence of edge subsets associated with (t_1, t_2). Then the minimal minor of G with respect to (t_1, t_2) coincides with A_Q.*

Proof Let M be the minimal minor of a graph G with respect to a maximally distant pair of trees (t_1, t_2) of G and let $e^* \in t_1^* \cap t_2^*$. From the equivalence of statements (1) and (1″) of Proposition 5.13, $t_1 \cap M$ and $t_2 \cap M$ are disjoint circuit-less subsets of M. By construction of the sequence A_Q^k, $k \geq 0$, it is clear that $A_Q^k \subseteq M$, for every k and consequently, $A_Q \subseteq M$. According to Corollary 5.21, A_Q is a minor of G with respect to (t_1, t_2). But M is a minimal minor of G with respect to (t_1, t_2) and hence $M \subseteq A_Q$. Thus, $A_Q \subseteq M$ and $M \subseteq A_Q$ which imply that $M = A_Q$.   □

The next proposition is a strengthened version of Proposition 5.23.

Proposition 5.25 *There is a unique edge subset M of G which is the minimal minor of G with respect to every maximally distant pair of trees.*

Proof Let M_{12} and M_{ab} be two minimal minors of G with respect to maximally distant pairs of trees (t_1, t_2) and (t_a, t_b) respectively. Because (t_1, t_2) and (t_a, t_b) are equidistant pairs of trees, we have

$$|t_a^* \cap t_b^*| = |t_1^* \cap t_2^*|.$$

On the other hand, from Lemma 5.19,

$$
\begin{aligned}
|t_a^* \cap t_b^*| &\geq |M_{12}| - |(t_a \cap M_{12}) \cup (t_b \cap M_{12})| \\
&\geq |M_{12}| - |(t_1 \cap M_{12}) \cup (t_2 \cap M_{12})| = |t_1^* \cap t_2^*|
\end{aligned}
$$

and hence the following relations hold:

$$|t_a^* \cap t_b^*| = |M_{12}| - |(t_a \cap M_{12}) \cup (t_b \cap M_{12})| \tag{1}$$

and

$$|(t_1 \cap M_{12}) \cup (t_2 \cap M_{12})| = |(t_a \cap M_{12}) \cup (t_b \cap M_{12})|. \tag{2}$$

Since $(t_a \cap M_{12}) \cup (t_b \cap M_{12}) \subseteq M_{12}$, (1) implies

$$|t_a^* \cap t_b^*| = |M_{12} \setminus (t_a \cap M_{12}) \cup (t_b \cap M_{12})|. \tag{3}$$

Also, for any subset M and for any pair of trees (t_a, t_b),

$$|t_a \cap M| + |t_b \cap M| \geq |(t_a \cap M) \cup (t_b \cap M)|, \tag{4}$$

where equality is attained iff $t_a \cap M$ and $t_b \cap M$ are disjoint. Because M_{12} is a minor of G with respect to the tree pair (t_1, t_2), from Proposition 5.13, $t_1 \cap M_{12}$ and $t_2 \cap M_{12}$ are disjoint subsets of M_{12} and consequently,

$$|(t_1 \cap M_{12}) \cup (t_2 \cap M_{12})| = |(t_1 \cap M_{12})| + |t_2 \cap M_{12}|.$$

But $t_1 \cap M_{12}$ and $t_2 \cap M_{12}$ are also maximal circuit-less subsets of M_{12} and hence

$$|t_1 \cap M_{12}| = |t_2 \cap M_{12}|, \tag{5}$$

$$|t_1 \cap M_{12}| \geq |t_a \cap M_{12}| \tag{6}$$

and

$$|t_2 \cap M_{12}| \geq |t_b \cap M_{12}|. \tag{7}$$

Accordingly,

$$|(t_1 \cap M_{12}) \cup (t_2 \cap M_{12})| = 2|(t_1 \cap M_{12})| \tag{8}$$

and

$$2|(t_1 \cap M_{12})| \geq |t_a \cap M_{12}| + |t_b \cap M_{12}|. \tag{9}$$

According to (4), for $M = M_{12}$ we have

$$|t_a \cap M_{12}| + |t_b \cap M_{12}| \geq |(t_a \cap M_{12}) \cup (t_b \cap M_{12})|. \tag{10}$$

Thus from (8), (9) and (10) we obtain:

$$|(t_1 \cap M_{12}) \cup (t_2 \cap M_{12})| \geq |t_a \cap M_{12}| + |t_b \cap M_{12}| \geq |(t_a \cap M_{12}) \cup (t_b \cap M_{12})|.$$

Combining (2), with the previous relation we conclude that

$$
\begin{aligned}
|(t_1 \cap M_{12}) \cup (t_2 \cap M_{12})| &= |t_a \cap M_{12}| + |t_b \cap M_{12}| \\
&= |(t_a \cap M_{12}) \cup (t_b \cap M_{12})|.
\end{aligned}
\tag{11}
$$

Hence, $t_a \cap M_{12})$ and $t_b \cap M_{12}$ are disjoint. Also, inserting (8) into (11) we obtain:

$$2|(t_1 \cap M_{12})| = |t_a \cap M_{12}| + |t_b \cap M_{12}|. \tag{12}$$

From (12), because of (6) and (7), we finally obtain:

$$|t_1 \cap M_{12}| = |t_a \cap M_{12}| = |t_b \cap M_{12}| = |t_2 \cap M_{12}|. \tag{13}$$

Thus, $t_a \cap M_{12}$ and $t_b \cap M_{12}$ are disjoint maximal circuit-less subsets of M_{12} and therefore no edge in $P(a,b) = t_a \cap t_b$ belongs to M_{12}. Hence, $F(a,b) = M_{12} \backslash ((t_a \cap M_{12}) \cup (t_b \cap M_{12}))$ is a subset of $Q(a,b) = t_a^* \cap t_b^*$. But $|F(a,b)| = |Q(a,b)|$, according to (3), and hence $F(a,b) = Q(a,b)$. On the other hand $F(a,b) \subseteq M_{12}$ and hence $Q(a,b) \subseteq M_{12}$. From the equivalence of (1) and (1'') (Proposition 5.13) M_{12} is a minor of G with respect to the tree pair (t_a, t_b). According to our assumption M_{ab} is unique minimal minor G with respect to the tree pair (t_a, t_b) and hence $M_{12} \subseteq M_{ab}$.

Repeating the above arguments, exchanging (t_1, t_2) with (t_a, t_b) we conclude that $M_{ab} \subseteq M_{12}$. Thus, $M_{12} \subseteq M_{ab}$ and $M_{ab} \subseteq M_{12}$ which imply that $M_{12} = M_{ab}$. $\square$

Thus, for any graph G, there is the unique edge subset M_o which is a minimal minor of G, with respect to every maximally distant pair of trees. This subset of edges of a graph G is called the *principal minor* of G with respect to maximally distant pairs of trees. According to Proposition 5.24, to find M_o it is sufficient to find the set A_Q with respect to a maximally distant pair of trees (t_1, t_2) of a graph. Recall that M_o contains all the edges in $t_1^* \cap t_2^*$ and none of the edges in $t_1 \cap t_2$.

We could again develop all the material of this section in an entirely dual manner substituting circuits for cutsets, A_Q for B_P and 'minor of G with respect to a pair of trees' for 'cominor of G with respect to the pair of trees'. We say that a cominor N of G with respect to a tree pair (t_1, t_2) of G is a *minimal cominor* of G with respect to this tree pair if no proper subset of N is also a cominor of G with respect to the same tree pair (t_1, t_2).

Consequently, the following dual statements hold:

Dual of Proposition 5.23 *There is a unique minimal cominor of a graph G with respect to a maximally distant pair of trees of G.*

Dual of Proposition 5.24 *Let (t_1, t_2) be a maximally distant pair of trees of a graph G and let B_P be the largest subset of the principal cosequence of edge subsets associated with (t_1, t_2). Then the minimal cominor of G with respect to (t_1, t_2) coincides with B_P.*

Dual of Proposition 5.25 *There is a unique edge subset N of G which is the minimal cominor of G with respect to every maximally distant pairs of trees.*

Thus, for any graph G, there is a unique edge subset N_o which is the minimal cominor of G with respect to every maximally distant pairs of trees. This subset of edges of a graph G is called the *principal cominor* of G with respect to a maximally distant pair of trees. According to the dual of Proposition 5.24, to find N_o it is sufficient to find the set B_P with respect to a maximally distant pair of trees (t_1, t_2) of G. N_o contains all the edges in $t_1 \cap t_2$ and none of the edges in $t_1^* \cap t_2^*$.

5.5   Hybrid pre-rank and the principal minor

Given a subset A of the edge set E of a graph G, let $n_s(A)$ (respectively, $n_c(A)$) denote the maximal number of independent cutsets (circuits) of G made from edges in A only. Recall that rank (A) (corank (A)) is defined to be a maximum circuit-less (cutset-less) subset of A. A set function $r_h(A)$, defined by $r_h(A) = \text{rank}\,(A) + \text{corank}\,(E \backslash A)$, we call the *hybrid pre-rank* of the subset A. Clearly, for every subset A of the edge set E of a graph G, $r_h(A)$ is a positive integer. The minimal value of the hybrid pre-rank $r_h(A)$ over all edge subsets A of G (that is, the number $r_o = \min_{A \subseteq E} r_h(A)$) is equal to the hybrid rank of G (Proposition 3.31, Chapter 3). In the literature devoted to the analysis of electrical circuits, the same number is called the topological degree of freedom. Note that $r_o \leq \min\{\text{rank}\,(G), \text{corank}\,(G)\}$.

Because corank $(E \backslash A) = \text{corank}\,(E) - |A| + \text{rank}\,(A)$ (Corollary 2.27, Chapter 2), we immediately obtain:

$$r_h(A) = \text{corank}\,(E) - g(A),$$

where $g(A) = |A| - 2\text{rank}\,(A)$. But rank $(A) = |A| - n_c(A)$, and therefore $g(A)$ can be rewritten in the following form: $g(A) = 2n_c(A) - |A|$.

It is obvious that $\min_{A \subseteq E} r_h(A) = \operatorname{corank}(E) - \max_{A \subseteq E} g(A)$. We denote $\max_{A \subseteq E} g(A)$ by g_m. Thus if g_m is known, r_o can be found from the relation:

$$r_o = \operatorname{corank}(E) - g_m.$$

Since $g(\emptyset) = 0$, it follows that $g_m \geq 0$.

Lemma 5.26 *For any two edge subsets A' and A'' of a graph G,*

$$g(A' \cup A'') \geq g(A') + g(A'') - g(A' \cap A'').$$

Proof Because

$$|A' \cup A''| = |A'| + |A''| - |A' \cap A''|$$

and

$$n_c(A' \cup A'') \geq n_c(A') + n_c(A'') - n_c(A' \cap A'')$$

it follows that,

$$
\begin{aligned}
g(A' \cup A'') &= 2n_c(A' \cup A'') - (|A'| + |A''| - |A' \cap A''|) \\
&\geq 2(n_c(A') + n_c(A'') - n_c(A' \cap A'')) \\
&\quad -(|A'| + |A''| - |A' \cap A''|) \\
&= (2n_c(A') - |A'|) + (2n_c(A'') - |A''|) \\
&\quad -(2n_c(A' \cap A'') - |A' \cap A''|) \\
&= g(A') + g(A'') - g(A' \cap A'').
\end{aligned}
$$

Thus, we finally obtain

$$g(A' \cup A'') \geq g(A') + g(A'') - g(A' \cap A''). \qquad \square$$

Lemma 5.27 *Let A' and A'' be two distinct edge subsets of a graph G such that $g(A') = g_m = g(A'')$. Then, $g(A' \cup A'') = g_m$ and $g(A' \cap A'') = g_m$.*

Proof Let $g(A') = g_m = g(A'')$. According to Lemma 5.26, $g(A' \cup A'') + g(A' \cap A'') \geq g(A') + g(A'')$ and hence $g(A' \cup A'') + g(A' \cap A'') \geq 2g_m$. Since $g(A) \leq g_m$ for every A, it follows that $g(A' \cup A'') \leq g_m$ and $g(A' \cap A'') \leq g_m$. Hence, $g(A' \cup A'') + g(A' \cap A'') \leq 2g_m$. Consequently, $g(A' \cup A'') + g(A' \cap A'') = 2g_m$. But $g(A' \cup A'') \leq g_m$ and $g(A' \cap A'') \leq g_m$ and therefore $g(A' \cup A'') = g_m$ and $g(A' \cap A'') = g_m$. $\qquad \square$

We say that A_o is a maximal subset of G such that $g(A_o) = max_{A \subseteq E} g(A)$ if there is no proper subset of A_o with the same property.

Proposition 5.28 *There is a unique minimal subset A_o of a graph G such that $g(A_o) = g_m = \max_{A \subseteq E} g(A)$.*

Proof Let A'_o be a minimal edge subset of G such that $g(A'_o) = g_m = \max_{A \subseteq E} g(A)$ and let A''_o be another minimal edge subset of G such that $g(A''_o) = g_m = \max_{A \subseteq E} g(A)$. According to Lemma 5.27, $g(A'_o \cap A''_o) = g_m = \max_{A \subseteq E} g(A)$. Since $A'_o \cap A''_o \subseteq A'_o$ and $A'_o \cap A''_o \subseteq A''_o$, and because of the minimality of subsets A'_o and A''_o we conclude that $A'_o \cap A''_o = A'_o$ and $A'_o \cap A''_o = A''_o$. Thus, $A'_o = A''_o$. $\qquad\square$

Lemma 5.29 *Let (t_1, t_2) be a pair of trees of a graph G and let F be an edge subset of G. Then $g(F) \le |F \cap t_1^* \cap t_2^*|$. Equality is attained iff $F \cap t_1$ and $F \cap t_2$ are disjoint maximal circuit-less subsets of F.*

Proof For every subset F of the edge set of a graph G:

$$\mathrm{rank}\,(F) \ge |F \cap t_1| \tag{1}$$

$$\mathrm{rank}\,(F) \ge |F \cap t_2| \tag{2}$$

and

$$|F| = |F \cap t_1| + |F \cap t_2| + |F \cap t_1^* \cap t_2^*| - |F \cap t_1 \cap t_2|. \tag{3}$$

Obviously, because $|F \cap t_1 \cap t_2| \ge 0$, it follows from (3) that

$$|F| \le |F \cap t_1| + |F \cap t_2| + |F \cap t_1^* \cap t_2^*| \tag{4}$$

while, (1)+(2) gives

$$2(\mathrm{rank}\,(F)) \ge |F \cap t_1| + |F \cap t_2|. \tag{5}$$

If we substitute (4) and (5) into the expression for $g(F)$ we obtain:

$$
\begin{aligned}
g(F) &= |F| - 2(\mathrm{rank}\,(F)) \\
&\le |F \cap t_1| + |F \cap t_2| + |F \cap t_1^* \cap t_2^*| - 2(\mathrm{rank}\,(F)) \\
&\le |F \cap t_1| + |F \cap t_2| + |F \cap t_1^* \cap t_2^*| - |F \cap t_1| - |F \cap t_2| \\
&= |F \cap t_1^* \cap t_2^*|.
\end{aligned}
$$

Thus finally,

$$g(F) \le |F \cap t_1^* \cap t_2^*|. \tag{6}$$

If we substitute (3) into the expression for $g(F)$ we obtain:

$$
\begin{aligned}
g(F) &= |F| - 2(\mathrm{rank}\,(F)) \\
&= |F \cap t_1| + |F \cap t_2| + |F \cap t_1^* \cap t_2^*| \\
&\quad - |F \cap t_1 \cap t_2| - 2(\mathrm{rank}\,(F)).
\end{aligned}
\tag{7}
$$

Comparing (6) and (7) we conclude that equality in (6) is attained iff

$$|F \cap t_1| + |F \cap t_2| - 2(\mathrm{rank}\,(F)) + (-|F \cap t_1 \cap t_2|) = 0. \tag{8}$$

Because, $|F \cap t_1| + |F \cap t_2| - 2(\text{rank}\,(F)) \leq 0$ (as a consequence of (5)) and because $(-|F \cap t_1 \cap t_2|) \leq 0$, it follows immediately that relation (8) holds iff $|F \cap t_1| + |F \cap t_2| - 2(\text{rank}\,(F)) = 0$ and, at the same time, $(-|F \cap t_1 \cap t_2|) = 0$. On the other hand, $|F \cap t_1| + |F \cap t_2| - 2(\text{rank}\,(F)) = 0$ iff $\text{rank}\,(F) = |F \cap t_1|$ and $\text{rank}\,(F) = |F \cap t_2|$. That is, iff $F \cap t_1$ and $F \cap t_2$ are maximal circuit-less subsets of F. Also, $|F \cap t_1 \cap t_2| = 0$ iff $F \cap t_1$ and $F \cap t_2$ are disjoint.

Thus relation (8) holds iff $F \cap t_1$ and $F \cap t_2$ are disjoint maximal circuit-less subsets of F. Hence under the same conditions equality in (6) is attained.   □

Corollary 5.30 [to Lemma 5.29] *An edge subset M of a graph G is a minor of G with respect to a maximally distant pair of trees (t_1, t_2) of G iff $g(M) = |t_1^* \cap t_2^*|$.*

Proof Let M be a minor of G with respect to the tree pair (t_1, t_2). Then, from Proposition 5.13, $M \cap t_1$ and $M \cap t_2$ are disjoint maximal circuit-less subsets of M and $t_1^* \cap t_2^* \subseteq M$. Hence, according to Lemma 5.29, $g(F) = |M \cap t_1^* \cap t_2^*|$. But $M \cap t_1^* \cap t_2^* = t_1^* \cap t_2^*$ and consequently, $g(M) = |t_1^* \cap t_2^*|$.

Conversely, let $g(M) = |t_1^* \cap t_2^*|$. Then, $|M \cap t_1^* \cap t_2^*| = |t_1^* \cap t_2^*|$ which implies that $t_1^* \cap t_2^* \subseteq M$. Also, according to Lemma 5.29, from $g(M) = |t_1^* \cap t_2^*|$ it follows that $M \cap t_1$ and $M \cap t_2$ are disjoint maximal circuit-less subsets of M. Hence, from Proposition 5.13, M is a minor of G with respect to the tree pair (t_1, t_2).   □

Proposition 5.31 *A subset M_o of the edge set E of a graph G is the principal minor of G with respect to maximally distant pairs of trees iff $g(M_o) = \max_{F \subseteq E} g(F)$ and no proper subset of M_o satisfies the same relation.*

Proof Let M_o be the principal minor of G with respect to maximally distant pairs of trees and let (t_1, t_2) be a maximally distant pair of trees of G. Because M_o is a minor of G with respect to every maximally distant pair of trees of G it is also a minor with respect to (t_1, t_2). Then, according to Corollary 5.30, $g(M_o) = |t_1^* \cap t_2^*|$. Let F be an arbitrary edge subset of G then, from Lemma 5.29, $g(F) \leq |F \cap t_1^* \cap t_2^*|$. But $|F \cap t_1^* \cap t_2^*| \leq |t_1^* \cap t_2^*|$, for any edge subset F and therefore, $g(F) \leq g(M_o)$, that is, $g(M_o) = \max_{F \subseteq E} g(F)$. To show that no proper subset M_1 of M_o satisfies the relation $g(M_1) = \max_{F \subseteq E} g(F)$, it is sufficient to show that $F \subset M_o$ implies $g(F) < g(M_o)$. Assume $F \subset M_o$. Since $M_o \cap t_1 \cap t_2$ is empty, it follows that $F \cap t_1 \cap t_2$ is also empty and hence $|F| = |F \cap t_1| + |F \cap t_2| + |F \cap t_1^* \cap t_2^*|$. Consequently,

$$
\begin{aligned}
g(F) &= |F| - 2(\text{rank}\,(F)) \\
&= (|F \cap t_1| + |F \cap t_2| - 2(\text{rank}\,(F))) + |F \cap t_1^* \cap t_2^*|.
\end{aligned}
\tag{1}
$$

Obviously, the following relations always hold:

$$|F \cap t_1^* \cap t_2^*| \leq |M_o \cap t_1^* \cap t_2^*| = |t_1^* \cap t_2^*| \tag{2}$$

$$|F \cap t_1| + |F \cap t_2| - 2(\text{rank}(F)) \leq 0. \tag{3}$$

We show that in at least one of (1) and (2) equality does not occur. Suppose that it is not true. That is, suppose $|F \cap t_1^* \cap t_2^*| = |t_1^* \cap t_2^*|$ and $|F \cap t_1| + |F \cap t_2| - 2(\text{rank}(F)) = 0$. According to (1), $g(F) = |t_1^* \cap t_2^*|$ and hence F is a minor of G with respect to the maximally distant pair of trees (t_1, t_2). But $F \subset M_o$ which contradicts the assumption that M_o is the minimal minor of G with respect to (t_1, t_2). Hence in at least one of (1) and (2) equality does not occur. This immediately implies that $g(F) < g(M_o)$. Thus, M_o is a minimal subset M_o satisfying $g(M_o) = \max_{F \subseteq E} g(F)$.

Conversely, suppose that there is a set A_o which is a minimal edge subset of G distinct from M_o satisfying $g(A_o) = \max_{A \subseteq E} g(F)$. Since M_o also satisfies the same relation (according to the *if part* of this proposition), we conclude that there are at least two minimal edge subset satisfying this relation. This contradicts the Proposition 5.28 and therefore $A_o = M_o$. □

Let r_o denote the hybrid rank of G, d_o the maximum tree pair distance over all tree pairs of G, b_o the maximum basoid cardinality over all basoids of G, and $r_h(M_o)$ the hybrid pre-rank of the principal minor M_o of G with respect to maximally distant pairs of trees. The following proposition connects these numbers.

Proposition 5.32 *For a graph G, the following relations hold: $r_o = d_o = b_o = r_h(M_o)$.*

Proof $(r_o = r_h(M_o))$: From, $r_o = \min_{F \subseteq E} r_h(F) = \text{corank}(E) - \max_{F \subseteq E} g(F)$, and $g(M_o) = |M_o| - 2(\text{rank}(M_o)) = \max_{F \subseteq E} g(F)$ it follows that $r_o = \text{corank}(E) - (|M_o| - 2(\text{rank}(M_o))) = r_h(M_o)$.

$(d_o = r_h(M_o))$: Let (t_1, t_2) be a maximally distant pair of trees of a graph G and let M_o be the principal minor of G with respect to maximally distant pairs of trees. Clearly, M_o contains $(t_1^* \cap t_2^*)$ and, from Proposition 5.13, $t_1 \cap M_o$ and $t_2 \cap M_o$ are disjoint maximal circuit-less subsets of M_o. Hence the following relations hold:

$$|M_o| - (|t_1 \cap M_o| + |t_2 \cap M_o|) = |(t_1^* \cap t_2^*)| \tag{1}$$

and

$$2(\text{rank}(M_o)) = |t_1 \cap M_o| + |t_2 \cap M_o|. \tag{2}$$

Combining (1) and (2) we obtain:

$$|M_o| - 2(\text{rank}(M_o)) = |M_o| - (|t_1 \cap M_o| + |t_2 \cap M_o|) = |(t_1^* \cap t_2^*)|. \tag{3}$$

Since $r_o = r_h(M_o)$ we have:

$$r_o = \text{corank}(E) - (|M_o| - 2(\text{rank}(M_o))). \tag{4}$$

Finally, substituting *(3)* into (4) we obtain:

$$r_o = \operatorname{corank}(E) - |(t_1^* \cap t_2^*)| = (|t_1 \backslash t_2| + |(t_1^* \cap t_2^*)|) - |(t_1^* \cap t_2^*)| = |t_1 \backslash t_2|. \quad (5)$$

Because (t_1, t_2) is a maximally distant pair of trees, it follows from (5) that the hybrid rank is equal to the maximal distance between two trees in G.

$(d_o = b_o)$: According to Proposition 5.8, a subset of edges b is a basoid of maximal cardinality iff b is the set difference of a maximally distant pair of trees. Hence $\max_G |b| = \max_G |(t_1 \backslash t_2)|$ over all basoids b and over all pairs of trees (t_1, t_2) of the graph.   □

Given a graph G, let M_o be the principal minor with respect to maximally distant pairs of trees and let N_o be the principal cominor of G with respect to maximally distant pairs of trees.

Proposition 5.33 *The principal minor and the principal cominor of a graph G with respect to maximally distant pairs of trees are disjoint.*

Proof Let (t_1, t_2) be a maximally distant pair of trees of a graph G. From Propositions 5.23, 5.24 and 5.25 it follows that $M_o = A_Q$ is the principal minor of G with respect to maximally distant pairs of trees. Similarly, from duals of Propositions 5.23, 5.24 and 5.25 it follows that $N_o = B_P$ is the principal cominor of G with respect to maximally distant pairs of trees. Since, (t_1, t_2) is a maximally distant pair of trees of a graph G, according to the implication $(1) \Rightarrow (4)$ of Proposition 5.20, A_Q and B_P are disjoint and therefore M_o and N_o are also disjoint.   □

According to Proposition 5.33, $M_o \cap N_o$ is empty and hence the pair (M_o, N_o) defines a 3-partition $(M_o, N_o, K_o = E \backslash (M_o \cup N_o))$ of G. This 3-partition is called the *principal partition* of a graph *with respect to maximally distant pairs of trees*. For the graph of Figure 5.5, the principal partition with respect to maximally distant pairs of trees is the triple $(M_o = \{8, 9, 10\}$, $N_o = \{1, 2, 3\}$, $K_o = \{4, 5, 6, 7\})$. The principal partition was first introduced for graphs in the context of tree pairs and then by other authors in a more general context.

5.6   Principal partition and Shannon's game

In this section we illustrate three facets of the principal partition of a graph. We first prove that the concept of the principal partition with respect to pairs of trees of maximum Hamming distance, introduced in this chapter, coincides with the concept of the principal partition with respect to dyads, introduced in Chapter 3. In order to prove this it is enough to show that the concept of the principal minor (respectively, cominor) of a graph with respect to dyads coincides with the concept of the principal minor (cominor) of a graph with respect to maximally distant pairs of trees.

Proposition 5.34 *A subset M_o of a graph G is the principal minor of G with respect to maximally distant tree pairs iff it is the principal minor of G with respect to dyads of G.*

Proof Let M'_o be the principal minor of G with respect to every dyad of G. Then, according to Proposition 3.14 of Chapter 3, $b \cap M'_o$ is a maximal circuit-less subset of M'_o and $b \cap (M'_o)^*$ is a maximal cutset-less subset of $(M'_o)^*$. From Proposition 5.8, for dyad b there exists a maximally distant pair of trees (t_1, t_2) so that $b = t_1 \setminus t_2$. Clearly, the set difference $t_2 \setminus t_1$ is also a dyad of G. Consequently, $(t_1 \setminus t_2) \cap M'_o$ is a maximal circuit-less subset of M'_o and $(t_2 \setminus t_1) \cap (M'_o)^*$ is a maximal cutset-less subset of $(M'_o)^*$. From Proposition 5.13 it follows that M'_o is a minor with respect to (t_1, t_2). If we repeat this procedure for all dyads, according to Proposition 5.8, all maximally distant pairs of trees will be involved. Therefore M'_o is a minor of G with respect to every maximally distant pair of trees of G.

Conversely, let M''_o be the principal minor of G with respect to maximally distant pairs of trees of G. Then, according to Proposition 5.13, for every maximally distant pair of trees (t_1, t_2), $(t_1 \setminus t_2) \cap M''_o$ is a maximal circuit-less subset of M''_o and $(t_2 \setminus t_1) \cap (M''_o)^*$ is a maximal cutset-less subset of $(M''_o)^*$. According to Proposition 5.8, the set difference $b = t_1 \setminus t_2$ is a dyad of G and hence $b \cap M''_o$ is a maximal circuit-less subset of M''_o and $b \cap (M''_o)^*$ is a maximal cutset-less subset of $(M''_o)^*$. Thus M''_o is a minor of G with respect to dyad b. If we repeat this procedure for all maximally distant pairs of trees of G, according to Proposition 5.8 all dyads will be involved. Therefore, according to Proposition 3.14 of Chapter 3, M''_o is a minor of G with respect to every dyad of G.

Because a minor with respect to a dyad is a subset of the principal minor M'_o with respect to dyads, it follows immediately that $M''_o \subseteq M'_o$. On the other hand a minor with respect to a maximally distant pair of trees is a subset of M''_o and therefore $M'_o \subseteq M''_o$. Thus, we have proved that $M'_o \subseteq M''_o$ and $M''_o \subseteq M'_o$ which imply that $M'_o = M''_o$. □

The concept of principal partition, although notionally introduced for the first time in the context of maximally distant pairs of trees, had been discovered earlier (implicitly) in connection with Shannon's famous switching game. The discovery was hidden within winning strategies. Shannon's switching game, which we will shortly describe, is a game played on a graph (or a matroid) with a single distinguished element e. The two players, Short and Cut, alternately choose elements of $E \setminus \{e\}$ with elements chosen by Cut deleted from the graph (matroid) and element chosen by Short contracted. The objective of Short is to collect a subset of elements of $E \setminus \{e\}$ which spans $\{e\}$

by means of circuits and the objective of Cut is to collect a subset of elements of $E \setminus \{e\}$ which spans $\{e\}$ by means of cutsets. Cut attempts to tag a subset of $E \setminus \{e\}$ which makes the objective of Short unattainable and vice versa. Every element of $E \setminus \{e\}$ can be chosen at most once, providing that both players have complete information. A game is called *short* (*cut*) if Short (Cut), playing second, can win against every strategy of Cut (Short). A game is *neutral* if the player going first can win against every strategy of the other player.

The next assertion shows that the set of edges of a graph can be divided into three disjoint subsets according to the outcome of the game played:

Assertion 5.35 *Let G be a graph with edge set E and let e be a distinguished element of E. Then for a game played on G with respect to e, exactly one of the following outcomes is possible:*

(1) *the game on G with respect to $e \in E$ is short.*

(2) *the game on G with respect to $e \in E$ is cut.*

(3) *the game on G with respect to $e \in E$ is neutral.*

Assertion 5.35 introduces a unique partition of the edge set of a graph into three subsets. Although related to the switching game, this partition does not inherently depend on the particular terminology of Shannon's switching game and coincides with the principal partition.

The next three assertions show that the partition of the edge set of a graph which corresponds to Shannon's switching game and the principal partition of the graph are related.

Assertion 5.36 *Let G be a graph with edge set E and let e be a distinguished element of E. Then the following statements are equivalent:*

(i) *A game on G with respect to $e \in E$ is a short game.*

(ii) *There are two disjoint circuit-less subsets I_1 and I_2 such that $e \in sp_c(I_1) = sp_c(I_1)$ and $e \notin I_1 \cup I_2$.*

(iii) *e belongs to the principal minor of G.*

Assertion 5.37 *Let G be a graph with edge set E and let e be a distinguished element of E. Then the following statements are equivalent:*

(i) *A game on G with respect to $e \in E$ is a cut game.*

(ii) *There are two disjoint cutset-less subsets J_1 and J_2 such that $e \in sp_s(J_1) = sp_s(J_1)$ and $e \notin J_1 \cup J_2$.*

(iii) *e belongs to the principal cominor of G.*

Assertion 5.38 *Let G be a graph with edge set E and let e be a distinguished element of E. Then the following statements are equivalent:*

(i) *A game on G with respect to $e \in E$ is a neutral game.*

(ii) *There are two disjoint circuit-less subsets I_1 and I_2 such that $sp_c(I_1) = sp_c(I_1)$ and either $e \in I_1$ or $e \in I_2$ and there are two disjoint cutset-less subsets J_1 and J_2 such that $sp_s(J_1) = sp_s(J_1)$ and either $e \in J_1$ or $e \in J_2$.*

(iii) *e belongs to the complement of the union of principal minor and principal cominor of G.*

5.7 Bibliographic notes

The concept of hybrid rank, the concepts of a maximally distant pair of trees and principal partition of a graph are probably the most valuable concepts in hybrid graph theory. These concepts, notionally introduced in the paper of Kishi and Kajitani [36], were subsequently considered in several papers and related to other concepts such as hybrid rank (Ohtsuki, Ishizaki and Watanabe [57]), hybrid tree graphs (Sengoku [63]), complementary trees (Lin [39]) as well as to perfect pairs of trees, superperfect pairs of trees and basoids (Novak and Gibbons [52, 54, 55, 56]). The important fact that the cardinality of a largest basoid of a graph is equal to the maximal tree diameter in the graph, was first pointed out by Sengoku [63] and recalled by Lin [39]. Topological degree of freedom (Ohtsuki, Ishizaki and Watanabe [57]) is another name for hybrid rank. This name occurs particularly in the literature concerned with the analysis of electrical circuits. Iri [31] provides analogous results for matrices to those of Kishi and Kajitani [36] through the introduction of the so-called minimum term rank of a matrix. The notions of principal partition and maximally distant pairs of trees have been also extended to matroids by Bruno and Weinberg [4] through the concept of principal minor. Some material of section 5.4 and 5.5 follows the paper of Bruno and Weinberg [4]. The concept of principal partition, although notionally introduced by Kishi and Kajitani ([36]) had been discovered earlier (implicitly) by Lehman [38] in connection with Shannon's famous switching game. The discovery was hidden within Lehman's winning strategies until Bruno and Weinberg's realization [4]. Lehman [38] generalized Shannon's game from graphs to matroids (graphoids) and solved it. The proofs of Assertions 5.36, 5.37 and 5.38 follow from the results described in [38] and [4].

Bibliography

[1] Abdullah, K., 'A necessary condition for complete solvability of RLCT networks', *IEEE Transactions on Circuit Theory*, **CT-19**, 492–493, (1972).

[2] Acketa, D., *The Catalogue of all Non-Isomorphic Matroids on at Most 8 Elements*, Special Issue, 1, Institute of Mathematics, Faculty of Sciences, Novi Sad, (1983).

[3] Amary, S., 'Topological foundation of Kron's tearing of electrical networks', *RAAG Memoirs*, **3**, 322, (1962).

[4] Bruno, J. and Weinberg, L., 'The principal minors of a matroid', *Linear Algebra and its Applications*, **4**, 17–54, (1971).

[5] Bryant, P.R. and Tow, J., 'The A-matrix of linear passive reciprocal networks', *Journal of the Franklin Institute*, **293**, 401–419, (1972).

[6] Bryant, V. and Perfect, V., *Independence Theory in Combinatorics*, Chapman and Hall, London, (1980).

[7] Bryton, R.K. and Moser, J.K., 'A theory of nonlinear networks', *Quarterly Applied Mathematics*, **22**, 1–33, (1964).

[8] Chen, W.-K., *Applied Graph Theory*, North-Holland, Amsterdam, (1971).

[9] Chen, W.-K., 'On vector spaces associated with a graph', *SIAM Journal of Applied Mathematics*, **20**, 526–529, (1971).

[10] Chen, W.-K., 'Recent advances in the application of graph theory to networks', *IEEE Circuits and Systems Magazine*, 12–21, December (1983).

[11] Chua, L.O. and Chen, L.-K., 'Diacoptic and generalized hybrid analysis', *IEEE Transactions on Circuits and Systems*, **CAS-23**, No 12, 694–705, (1976).

[12] Chua, L.O. and Lin, P.M., *Computer-Aided Analysis of Electronic Circuits: Algorithms and Computational Techniques*, Prentice Hall, Englewood Cliffs, NJ, (1975).

[13] Chua, L.O. and McPherson, J.D., 'Explicit topological formulation of Lagrangian and Hamiltonian equations for nonlinear networks', *IEEE Transactions on Circuits and Systems*, **CAS-21**, No 2, 277–286, (1974).

[14] Deo, N., *Graph Theory with Applications to Engineering and Computer Science*, Prentice-Hall, Englewood Cliffs, NJ, (1974).

[15] Deo N., 'A Central tree', *IEEE Transactions on Circuit Theory*, **CT-13**, 439–440 (1973).

[16] Desoer, C.A. and Kuh, E.S., *Basic Circuit Theory*, McGraw-Hill, New York (1969).

[17] Fosseprez, M., *Topologie et comportement des circuits non lineaires non reciproques*, Collection META, Presses Polytechniques Romandes, (1989).

[18] Fournier, J.C., 'Binary Matroids'. In: *Combinatorial Geometries*, edited by Neil White, Cambridge University Press, Cambridge, pp. 28–39, (1987).

[19] Gao, S. and Chen, W.-K., 'The hybrid method of network analysis and topological degree of freedom', *Proceedings of IEEE International Symposium on Circuits and Systems*, Rome, pp. 158–161, (1982).

[20] Garey, M.R. and Johnson, D.S., *Computers and Intractability: A Guide to the Theory of NP-Completeness*, W.H. Freeman, San Francisco, (1979).

[21] Gibbons, A., *Algorithmic Graph Theory*, Cambridge University Press, London, (1985).

[22] Goetschel, R. and Voxman, W., 'Fuzzy matroids', *Fuzzy Sets and Systems*, **27**, 291–302, (1988).

[23] Goetschel, R. and Voxman, W. 'Bases of fuzzy matroids', *Fuzzy Sets and Systems*, **31**, 253–261, (1989).

[24] Gould, R., 'Graphs and vector spaces', *Journal of Mathematical Physics*, **37**, 193–214, (1958).

[25] Hakimi, S.L., 'On trees of a graph and their generation', *Journal of the Franklin Institute*, **272**, 347–359, (1961).

[26] Harary, F., *Graph Theory*, Addison-Wesley, Reading, MA, (1969).

[27] Harary, F., Mokken, R.J., and Plantholt, M.J., 'Interpolation theorem for diameters of spanning trees', *IEEE Transactions on Circuits and Systems*, **CAS-30**, 429–431, (1983).

[28] Harary, F. and Welsh, D.J.A., 'Matroids versus graphs, the many facets of graph theory', *Springer Lecture Notes*, **110**, 155–170, (1969).

[29] Hasler, M., 'Non-linear non-reciprocal resistive circuits with a structurally unique solution', *International Journal of Circuit Theory and Applications*, **14**, 237–262, (1986).

[30] Hasler, M. and Neirynck, J., *Nonlinear Circuits*, Artech House, Norwood MA, (1986).

[31] Iri, M., 'The maximum-rank minimum-term-rank theorem for the pivotal transforms of a matrix', *Linear Algebra*, **2**, 427–446, (1969).

[32] Iri, M., 'Combinatorial cannonical form of a matrix with applications to the principal partition of a graph', *Transactions of the Institute of Electronics and Communication Engineers of Japan* (IECEJ), **54A**, 30–37, (1971).

[33] Iri, M., 'A review of recent work in Japan on principal partitions of matroids and their applications', *Annals of the New York Academy of Science*, **319**, 306–319, (1979).

[34] Iri, M. and Fujishige, S., 'Use of matroid theory in operations research, circuits and system theory', *International Journal Systems Science*, **12**, No.1, 27–54, (1981).

[35] Kirchhoff, G., 'Über die Auflösung der Gleichungen, auf welche man bei der Utersuchung der linearen Vertheilung galvanischer Ströme geführt wird', *Poggendorffs Ann. Phys. Chem.*, **72**, 497–508, (1847).

[36] Kishi, G. and Kajitani, Y., 'Maximally distant trees and principal partition of a linear graph', *IEEE Transactions on Circuit Theory*, **CT-16**, 323–330, (1969).

[37] Lawler, E.L., *Combinatorial Optimization: Networks and Matroids*, Holt, Rinehart and Winston, New York, (1976).

[38] Lehman, A., 'A solution to the Shannon switching game', *Journal of the Society of Industrial and Applied Mathematics*, **12**, No.4, 687–725, (1964).

[39] Lin, P.M., 'Complementary trees in circuit theory', *IEEE Transactions on Circuits and Systems*, **CAS-27**, 921–928, (1980).

[40] Mathis, W., *Theorie nichtlinearer Netzwerke*, Springer-Verlag, Berlin, (1987).

[41] Maxwell, L.M. and Reed, M.B., 'Subgraph identification - segs, circuits and paths', *Proceedings of the Midwest Symposium on Circuit Theory*, **8**, Fort Collins (1965).

[42] Mayeda, W., *Graph Theory*, John Wiley and Sons, Inc. New York, London, (1972).

[43] Milic, M.M., 'General passive networks – solvability, degeneracies and order of complexity', *IEEE Transactions on Circuits and Systems*, **CAS-21**, 177–183, (1974).

[44] Milic, M.M. and Novak, L.A., 'Formulation of equations in terms of scalar functions for lumped nonlinear networks', *International Journal of Circuit Theory and Applications*, **9**, 15–32, (1981).

[45] Milic, M.M. and Novak, L.A., 'On weak completeness and nondegeneracy of networks', *International Journal of Circuit Theory and Applications*, **17**, 409–419, (1989).

[46] Minty, G.J., 'On the axiomatic foundations of the theories of direct linear graphs, electrical networks and network programming', *Journal of Mathematical Mechanics*, **15**, 485–520, (1966).

[47] Newcomb, R.W., *Linear Multiport Synthesis*, McGraw-Hill, New York, (1966).

[48] Novak, L.A., 'Affine abstract circuits and representation theorem', *Proceedings of the 6th International Symposium on Networks, Systems and Signal Processing*, Zagreb, Faculty of Electrical Engineering, University of Zagreb, 290–292, (1989).

[49] Novak, L.A., 'A few theorems concerning pairs of matroid bases', *International Journal of Circuit Theory and Applications*, **18**, 205–208, (1990).

[50] Novak, L.A., 'Reciprocity and antireciprocity of resistive N-ports', *IEEE Transactions on Circuits and Systems*, **CAS-37**, 860–863, (1990).

[51] Novak, L.A., 'On fuzzy independence set systems', *Fuzzy Sets and Systems*, **91**, 365–374, (1997).

[52] Novak, L.A. and Gibbons, A.M., 'Perfect pairs of trees in graphs', *International Journal of Circuit Theory and Applications*, **20**, 201–208, (1992).

[53] Novak, L.A. and Gibbons, A.M., 'Perfect pairs of trees associated with a prescribed tree: an algorithmic approach', *International Journal of Circuit Theory and Applications*, **20**, 681–688, (1992).

[54] Novak, L.A. and Gibbons, A.M., 'Superperfect pairs of trees in graphs', *International Journal of Circuit Theory and Applications*, **21**, 183–189, (1993).

[55] Novak, L.A. and Gibbons, A.M., 'Basoids and Principal partition of a graph', *Research Report of the Department of Computer Science*, University of Warwick, (1993).

[56] Novak, L.A. and Gibbons, A.M., 'Double independence and tree pairs in graphs', *International Journal of Circuit Theory and Applications*, **24**, 657–666, (1996).

[57] Ohtsuki, T., Ishizaki, Y. and Watanabe H., 'Topological degrees of freedom and mixed analysis of electrical networks', *IEEE Transactions on Circuit Theory*, **CT-17**, 491–499, (1970).

[58] Ohtsuki, T. and Watanabe, H., 'State-variable analysis of RLC networks containing non-linear coupling elements', *IEEE Transactions on Circuit Theory*, **CT-16**, 26–38, (1969).

[59] Ozawa, T., 'Topological conditions for the solvability of linear active networks', *International Journal of Circuit Theory and Applications*, **4**, 125–136, (1976).

[60] Papadimitriou, C.H. and Steiglitz, K., *Combinatorial Optimization: Algorithms and Complexity*, Prentice-Hall Inc., Englewood Cliffs, NJ, (1982).

[61] Recski, A., *Matroid Theory and its Applications in Electric Network Theory and Statics*, Akademiai, Budapest, (1989).

[62] Reed, M.B., 'The seg: a new class of subgraphs', *IEEE Transactions on Circuit Theory*, **CT-8**, 17–22, (1961).

[63] Sengoku, M., 'Hybrid trees and hybrid tree graphs', *IEEE Transactions on Circuits and Systems*, **CAS-22**, 786–790, (1975).

[64] Sillamaa, H.V., *Fundamentals of Topological Synthesis of Interconnection Structure of Multiterminal Network Elements*, Valgus, Tallin, (1983).

[65] Swamy, M.N.S. and Thulasiraman, K., *Graphs, Networks, and Algorithms*, John Wiley & Sons, New York, (1981).

[66] Tomizawa, N., 'Strongly irreducible matroids and principal partition of a matroid into strongly irreducible minors', *Transaction of the Institute of Electronics and Communication Engineers of Japan* (IECEJ), **59A**, 83–91, (1976).

[67] Tsuchiya, T., Ohtsuki, T., Ishizaki, Y., Watanabe H., Kajitani, Y. and Kishi, G., 'Topological degrees of freedom of electrical networks', *Fifth Annual Atherton Conference on Circuit and System Theory*, (1967).

[68] Tutte, W.T., 'Lectures on matroids', *J. Res. Nat. Bur. Stand.*, 1 and 2, **69B**, (1965).

[69] Tutte, W.T., *Introduction to the Theory of Matroids*, American Elsevier, New York, (1971).

[70] Vandewalle, J. and Chua, L.O., 'The colored branch theorem and its applications in circuit theory', *IEEE Transactions of Circuits and Systems*, **CAS-27**, 816–825, (1980).

[71] Watanabe H., 'A computational method for network topology', *IEEE Transactions on Circuit Theory*, **CT-7**, 296–302, (1960).

[72] Weinberg, L., 'Matroids, generalized networks, and electrical network synthesis', *Journal of Combinatorial Theory, Series B*, 23, 106–126 (1977).

[73] Welsh, D.J.A., *Matroid Theory*, London Mathematics Society Monographs, No.8, Academic Press, London, (1976).

[74] Whitney, H., '2-isomorphic graphs', *American Journal of Mathematics*, **55**, 245–254, (1933).

[75] Whitney, H., 'On the abstract properties of linear dependence', *American Journal of Mathematics*, **57**, 509–533, (1935).

[76] Williams, T.W. and Maxwell, L.M., 'The decomposition of a graph and the introduction of new class of subgraphs', *SIAM Journal of Applied Mathematics*, **20**, 385–389, (1971).

[77] Wilson, R.J., *Introduction to Graph Theory*, Oliver and Boyd, Edinburgh, (1972).

Index

abstract network, 134
  affine, 134
  noncontradictory, 134

basic pair, 104
  optimal, 108
basoid, 82
binary graphoid, 40
  corank, 67
  dual, 67
  edge, 40
  graphic, 63, 68
  ground set, 40
  orientable, 73
  rank, 62
binary matroid
  circuit, 40
  cutset, 40
  nongraphic, 64

CCCS, 31
CCVS, 30
circ, 7, 40
  space, 11
circ basis, 67
  2-complete, 67
  extended, 67
circuit, 7, 40
  fundamental, 52
  oriented, 33
circuit-separator, 16
  elementary, 17
cominor
  with respect to a basoid, 91
  with respect to a pair of trees,
    145

component
  1-connected, 5
  2-connected, 18
  circuit-connected, 17
  cutset-connected, 17
corank, 48
cotree, 80, 115
coupling, 25
current, 27
current-controlled current source, 31
current controlled voltage source, 30
cut, 5, 40
  space, 10
cut basis, 62
  extended, 63
  extended 2-complete, 63, 67
cutset, 6, 40
  fundamental, 53
  oriented, 34
cutset-separator, 17
  elementary, 17

double independent set, 81
dyad, 82, 114

edge, 2
  directed, 23
  hyper, 2
  regular, 2
electrical network, 27
  oriented, 27

frame of reference, 25
fundamental circuit, 49
fundamental cutset, 49
fundamental exchange theorem, 50

graph, 1
- 1-connected (connected), 5
- 1-isomorphism, 19
- 2-connected, 7
- 2-isomorphism, 19
- bipartite, 8
- circuit-connected, 17
- connected, 5
- contraction, 3
- cutset-connected, 17
- directed (digraph), 23
- Eulerian (co-bipartite), 8
- partially directed, 24
- reduction, 3
- unconnected, 5

graphoid
- contraction, 42
- reduction, 42
- regular, 73

hybrid rank, 102, 114
hypergraph, 2

ideal gyrator, 30
ideal transformer, 30
incidence matrix
- circuit, 36
- cutset, 36
- fundamental circuit, 58
- fundamental cutset, 58
independent source
- current, 29
- voltage, 29

Kirchhoff's space, 38

loop-edge, 2

matroid
- circuit matroid, 41
- cographic, 41
- cutset matroid, 41
- graphic, 41
matroid base, 69
matroid cobase, 70

minor
- with respect to a basoid, 91
- with respect to a pair of trees, 145
multiport, 26
- global, 37

n-port
- affine, 28
- constitutive relation, 27
- linear, 28
- nonlinear, 28
- parallel to an affine n-port, 28
- regular, 29
- resistive, 28
- singular, 29
- time-invariant, 28
n-port, 26
network analysis
- hybrid, 105
- mesh, 76, 105
- nodal, 76, 105
network connection
- galvanic, 25
- non-galvanic (coupling), 25
network equations
- hybrid, 79, 108
- mesh, 79
- nodal, 78
norator, 29
nullator, 29
nulor, 30

Ohm's resisitor, 29
open-circuit, 29
operational amplifier, 30
orthogonality theorem, 12

Painting Theorem
- for partially directed graphs, 24
painting theorem
- for nondirected graphs, 14
pair of trees
- complementary, 144

disjoint, 144
   maximally distant, 141
   perfect, 121
   superperfect, 130
parallel edges, 2
   directed, 23
principal cominor
   with respect to a basoid, 102
   with respect to a pair of trees,
      159
principal cosequence
   with respect to a basoid, 95
   with respect to a pair of trees,
      149
principal minor
   with respect to a basoid, 101
   with respect to a pair of trees,
      158
principal partition
   with respect to a basoid, 102
   with respect to a pair of trees,
      164
principal sequence
   with respect to a basoid, 94
   with respect to a pair of trees,
      149

rank, 48
reduced incidence matrix
   circuit, 77
   cutset, 77
ring sum, 9

self-circuit, 7

self-cutset, 6
Shannon's switching game, 165
short circuit, 29
signal, 27
span, 81
star, 7
   separable, 7
subgraph, 3
   edge-induced, 4
   vertex-induced, 3

Tellegen's theorem, 38
topological relations, 34
topologically complete set of vari-
     ables, 106
tree, 80, 115
   central, 115
   diameter, 115
   extremal, 115
   normal, 139

VCCS, 30
VCVS, 30
vertex, 1
   degree, 2
   in-degree, 23
   isolated, 2, 23
   out-degree, 23
vertex-separator, 4
   elementary, 4
voltage, 27
voltage-controlled current source, 30
voltage controlled voltage source, 30